T0306008

Photovoltaic Materials and Manufacturing Issues

MATERIALS RESEARCH SOCIETY
SYMPOSIUM PROCEEDINGS VOLUME 1123

Photovoltaic Materials and Manufacturing Issues

Symposium held December 2–4, 2008, Boston, Massachusetts, U.S.A.

EDITORS:

Bhushan Sopori
National Renewable Energy Laboratory
Golden, Colorado, U.S.A.

Jeff Yang
United Solar Ovonic LLC
Troy, Michigan, U.S.A.

Thomas Surek
Surek PV Consulting
Denver, Colorado, U.S.A.

Bernhard Dimmler
Würth Solar GmbH & Co. KG
Schwäbisch Hall, Germany

Materials Research Society
Warrendale, Pennsylvania

CAMBRIDGE
UNIVERSITY PRESS

University Printing House, Cambridge CB2 8BS, United Kingdom

One Liberty Plaza, 20th Floor, New York, NY 10006, USA

477 Williamstown Road, Port Melbourne, VIC 3207, Australia

314-321, 3rd Floor, Plot 3, Splendor Forum, Jasola District Centre, New Delhi - 110025, India

79 Anson Road, #06-04/06, Singapore 079906

Cambridge University Press is part of the University of Cambridge.

It furthers the University's mission by disseminating knowledge in the pursuit of education, learning and research at the highest international levels of excellence.

www.cambridge.org
Information on this title: www.cambridge.org/9781605110950

Materials Research Society
506 Keystone Drive, Warrendale, PA 15086
http://www.mrs.org

© Materials Research Society 2009

First published 2009
First paperback edition 2012

Single article reprints from this publication are available through University Microfilms Inc., 300 North Zeeb Road, Ann Arbor, MI 48106

CODEN: MRSPDH

A catalogue record for this publication is available from the British Library

ISBN 978-1-605-11095-0 Hardback
ISBN 978-1-107-40846-3 Paperback

*Invited Paper

THIN FILM POLYCRYSTALLINE
MATERIALS AND DEVICES

*Invited Paper

CHARACTERIZATION

*Invited Paper

NEW TECHNOLOGIES

PREFACE

This volume is a collection of many of the papers presented at Symposium P, "Photovoltaic Materials and Manufacturing Issues," held December 2–4 at the 2008 MRS Fall Meeting in Boston, Massachusetts. The intent of this symposium was to provide a forum to discuss major issues pertaining to current and emerging materials for photovoltaic applications and those related to low-cost production of solar cells and modules.

Rapid technical advances in solar photovoltaic (PV) technologies have resulted in reduced costs, and improved conversion efficiencies and reliabilities for solar systems, resulting in rapidly growing markets through 2008. Materials research advances have been the basis for much of the rapid progress in PV technologies over the past three decades. This symposium attracted a cross-section of current industry research leaders, new start-up companies, and university and laboratory researchers. The talks reviewed the most significant advances in PV materials and devices research, and examined the research challenges to reach the ultimate potential of current-generation (crystalline silicon), next-generation (thin films and concentrators), and future-generation PV technologies. The latter include innovative materials and device concepts that hold the promise of significantly higher conversion efficiencies and/or much lower costs. The program was comprised of some 58 oral presentations (about one-third from overseas) and 22 posters (about two-thirds from overseas), but many papers had global collaborators from industry, universities, and laboratories.

These proceedings capture a cross section of the oral and poster presentations at the conference and provide a flavor of the progress and ongoing research activities. The organization of this proceedings volume follows general grouping of technologies based on silicon, thin film polycrystalline materials, characterization, and new approaches. We wish to thank all authors and attendees of the symposium for very stimulating presentations and the discussions that followed. We are grateful to the National Renewable Energy Laboratory for their support of the symposium. We also wish to thank the Meeting Chairs, all abstract and proceedings paper reviewers, and the MRS staff for their help.

Bhushan Sopori
Jeff Yang
Thomas Surek
Bernhard Dimmler

September 2009

MATERIALS RESEARCH SOCIETY SYMPOSIUM PROCEEDINGS

MATERIALS RESEARCH SOCIETY SYMPOSIUM PROCEEDINGS

Prior Materials Research Society Symposium Proceedings available by contacting Materials Research Society

Silicon Solar Cells and Devices

Mater. Res. Soc. Symp. Proc. Vol. 1123 © 2009 Materials Research Society   1123-P03-01

High-Efficiency HIT Solar Cells for Excellent Power Generating Properties

Toshihiro Kinoshita, Daisuke Ide, Yasufumi Tsunomura, Shigeharu Taira, Toshiaki Baba, Yukihiro Yoshimine, Mikio Taguchi, Hiroshi Kanno, Hitoshi Sakata and Eiji Maruyama
Advanced Energy Research Center, Sanyo Electric Co., Ltd, Kobe 651-2242, Japan

ABSTRACT

In order to achieve the widespread use of HIT (Hetero-junction with Intrinsic Thin-layer) solar cells, it is important to reduce the power generating cost. There are three main approaches for reducing this cost: raising the conversion efficiency of the HIT cell, using a thinner wafer to reduce the wafer cost, and raising the open circuit voltage to obtain a better temperature coefficient. With the first approach, we have achieved the highest conversion efficiency values of 22.3%, confirmed by AIST, in a HIT solar cell. This cell has an open circuit voltage of 0.725 V, a short circuit current density of 38.9 mA/cm^2 and a fill factor of 0.791, with a cell size of 100.5 cm^2. The second approach is to use thinner Si wafers. The shortage of Si feedstock and the strong requirement of a lower sales price make it necessary for solar cell manufacturers to reduce their production cost. The wafer cost is an especially dominant factor in the production cost. In order to provide low-priced, high-quality solar cells, we are trying to use thinner wafers. We obtained a conversion efficiency of 21.4% (measured by Sanyo) for a HIT solar cell with a thickness of 85μm. Even better, there was absolutely no sagging in our HIT solar cell because of its symmetrical structure. The third approach is to raise the open circuit voltage. We obtained a remarkably higher Voc of 0.739 V with the thinner cell mentioned above because of its low surface recombination velocity. The high Voc results in good temperature properties, which allow it to generate a large amount of electricity at high temperatures.

INTRODUCTION

To cope with increasing demands for high-quality solar cells from all over the world, we plan to expand the annual production of HIT solar cells from 340 MW in FY 2008 to more than 600 MW in FY 2010. At the same time, we will increase the pace of development to offer technical advantages into our future products. We plan to raise cell conversion efficiency to 23.5% in the laboratory by 2010.

We developed a high-efficiency solar cell structure known as the HIT structure, and have been raising its quality. We recently updated the world's highest conversion efficiency of 22.3 % with a practical-sized solar cell in June 2007 [1]. Also, using high-efficiency HIT cells, we have achieved a record conversion efficiency of 20.6% for an R&D prototype module, which was certified by Advanced Industrial Science and Technology (AIST) [2]. These technologies are being steadily transferred to production. This paper describes the high-efficiency technologies and characteristics of HIT solar cells.

HIT SOLAR CELL STRUCTURE

The HIT solar cell is primarily characterized by its high conversion efficiency. It is also appreciated for its excellent temperature coefficient, the potential that it offers for using a very thin Si wafer, and the possibility for its application to bifacial solar modules. As shown in figure 1, an intrinsic (i-type) amorphous silicon (a-Si) layer followed by a p-type a-Si layer are deposited on a randomly textured n-type CZ crystalline silicon (c-Si) wafer to form a p/n heterojunction. On the opposite side of the c-Si wafer, i-type and n-type a-Si layers are deposited to obtain a Back Surface Field (BSF) structure. On both sides of the doped a-Si layers, Transparent Conducting Oxide (TCO) layers as an antireflective coating consisting and metal grid electrodes are formed. The symmetrical structure of the HIT cell is suitable for a bifacial module. All processes are conducted in temperature below 200 °C.

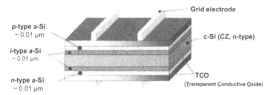

Figure 1. The structure of a HIT solar cell.

By developing a high-quality i-type a-Si layer, the defect level on the c-Si surface can be considerably reduced, and a high *Voc* can be obtained [3]. The excellent c-Si/a-Si hetero interface of the HIT structure leads to a outstandingly high *Voc* of more than 0.72 V. A higher *Voc* allows for not only a high conversion efficiency but also en excellent temperature coefficient, that excels the temperature coefficient of a diffused c-Si solar cell and is almost comparable to that of an amorphous Si solar cell. This feature results in higher output power even at higher temperatures.

CONVERSION EFFICIENCY TREND

Figure 2 shows the trend in the conversion efficiency of HIT solar cells. We are aiming for a high conversion efficiency of over 23.5% in our R&D by the end of 2010. As a result of our aggressive studies, we achieved a record high conversion efficiency of 22.3% (*Voc*: 0.725 V, *Isc*: 38.9 mA/cm^2, *FF*: 0.791, total area: 100.5 cm^2, certified by AIST) in July 2007, as shown in figure 3.

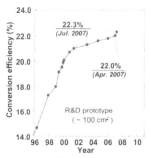

Figure 2. Trend in the conversion efficiency of HIT cells.

4

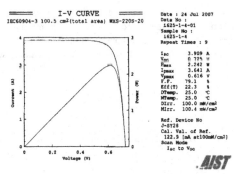

Figure 3. *I-V* characteristics of the HIT solar cell certified by AIST, which shows the world's highest conversion efficiency of 22.3% for a practical size cell (> 100 cm²).

As an approach for a product with even higher quality, we have been developing module technologies for maximizing long-term stability and module efficiency by taking advantage of the HIT cell's excellent features. As a result, we have demonstrated very high conversion efficiency with an R&D prototype module. The module efficiency has reached 20.6 %, which was certified by AIST in December 2007. This achievement also indicates the future superiority of HIT modules.

APPROACHES FOR HIGHER CONVERSION EFFICIENCY

We have been focusing on the following techniques for obtaining a high conversion efficiency with the HIT structure: (a) Improving the HIT structure by enhancing the a-Si/c-Si heterojunction properties for a higher *Voc*, (b) Improving the grid electrode for a higher *Isc* and *FF*, and (c) Reducing the absorption loss of the TCO layer for a higher *Isc*.

(a) Improving the HIT structure
The high *Voc* of the HIT solar cell is achieved by the effective passivation of c-Si surface defects with a high-quality intrinsic a-Si layer. The following fabrication processes are being used in the development of these solar cells:
HCleaning the c-Si surface with low-cost wet cleaning processes before a-Si deposition
HDepositing a high-quality intrinsic a-Si layer by chemical vapor deposition
HMaintaining low plasma and thermal damage to the c-Si surface and heterojunction while fabricating the a-Si, TCO layers and grid electrode.
By using these processes, the carrier recombination is decreased, resulting in fewer localized states in the intrinsic layer and interface of the heterojunction [4]. We have decided to target a *Voc* value of more than 0.74 V by unifying these techniques to achieve our conversion efficiency goal.

(b) Optimizing the grid electrode
For a higher *Isc* and *FF*, the grid electrode requires lower resistance and finer lines for a larger aperture, simultaneously. Since the grid electrode of HIT solar cells is made of silver (Ag)

5

paste, which has high resistivity in itself, the aspect ratio must be as high as possible. Figure 4(a) shows a conventional grid electrode fabricated by the screen-printing method. The conventional grid electrode has a spreading area that causes optical loss. In order to lower the optical loss and the resistance loss, it is necessary to eliminate this spreading area and raise the height, as shown in figure 4(b).

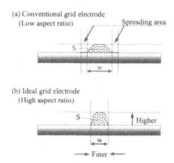

Figure 4. Schematic diagrams of (a) a conventional grid electrode with a spreading area and low aspect ratio, and (b) an ideal grid electrode with no spreading area and high aspect ratio.

(c) Reducing the absorption loss

There are optical losses in the short and long wavelength regions. The optical loss in the long wavelength region is mainly caused by the absorption of TCO. The HIT structure uses the TCO on the surface to collect the generating carriers. A lower level of optical absorption loss and a higher electrical conductivity for the TCO layer will lead to a higher Isc and FF. The optical absorption loss in the TCO is mainly caused by free carrier absorption. The deposition conditions of the TCO have been optimized to obtain a high-quality TCO layer with high carrier mobility. The new TCO shows better sensitivity in the long wavelength region (>1,000 nm) of the IQE spectra. Furthermore, the high FF value (0.791) in HIT solar cells using the new TCO suggests high conductivity for high carrier mobility [5].

UTILIZING A THIN WAFER FOR THE HIT CELL

The wafer accounts for about half the cost of a solar module. We have been developing thinner HIT solar cells to reduce the power-generating cost. Here, we have evaluated the possibility of thinner HIT solar cells for the first time. Figure 5 shows a picture of a HIT solar cell with a c-Si wafer thickness of ~85μm.

Figure 5. A HIT solar cell using an 85-μm-thick c-Si wafer. Note that no sagging is seen.

6

The *I-V* curve of the HIT solar cell on an 85-μm- thick c-Si wafer is shown in figure 6. We obtained a high conversion efficiency of 21.4% with this HIT solar cell. The *Voc* is an extremely high value of 0.739 V, the *Isc* is 37.3 mA/cm^2, and the *FF* is 77.6% [6]. These values were measured under standard conditions in-house. The *Jsc* value with the 85-μm-thick wafer decreased about 3% in comparison with that of a 165-μm-thick wafer. Those values corresponded approximately to the values calculated using SUNRAYS [7].

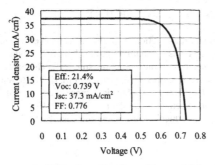

Figure 6. The *I-V* curve of a HIT solar cell with an 85-μm- thick c-Si wafer, which was measured at SANYO (measurement conditions: AM 1.5, 1 SUN, 25.0 °C, cell size (aperture size): 103.3 cm^2).

Figure 7 shows the thickness dependence of *I-V* parameters normalized by those of a 165-μm-thick HIT solar cell. As the thickness of the c-Si wafer decreases, the *Isc* decreases. This decrease in *Isc* is caused by insufficient light absorption and agrees well with the calculated result using SUNRAYS. On the other hand, the *Voc* shows no decrease because of the excellent HIT passivation effect on the c-Si surface. However, there seems to be a slight decrease in *FF* of about 0.5-1%. This result suggests that we still have room to improve the a-Si/c-Si heterojunction performance toward a thinner, next-generation c-Si wafer structure below 100μm.

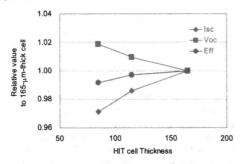

Figure 7. The *Isc*, *Voc* and conversion efficiency of HIT solar cells as functions of various HIT cell thicknesses. The values are normalized by that of a 165-μm-thick HIT cell.

7

We are now focusing on developing optical confinement technologies to improve the *Isc*, and enhanced passivation technologies to improve the *FF* toward the next generation. These results suggest that the HIT structure is suitable for thinner Si wafers with high conversion efficiency. However, the light confinement design is a very important factor for further improving the *Isc* in a thin HIT cell structure. This will enable the HIT structure to reduce the wafer cost while maintaining high conversion efficiency.

IMPROVING THE TEMPERATURE COEFFICIENT

Generally, an excellent temperature coefficient for the conversion efficiency enables high performance in outdoor use, which leads to a number of user benefits. As shown in figure 8, the temperature coefficient mainly depends on the *Voc* of the solar cell. Therefore, the HIT solar cell with its high *Voc* exhibits an excellent temperature coefficient [8,9]. The temperature coefficient of the HIT solar cell with 0.736 *V* is -0.23 %/°C. This is comparable to that of an a-Si solar cell, which is generally said to have an excellent temperature coefficient. Consequently, improving the temperature coefficient helps to reduce power-generating costs by further improving the *Voc*, such as by thinning wafers.

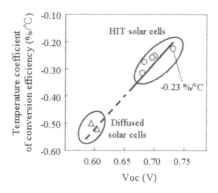

Figure 8. Temperature coefficient of conversion efficiencies as a function of the *Voc*s of solar cells.

APPLICATION TO A BIFACIAL SOLAR MODULE

The front and back symmetrical structure of the HIT solar cell, as shown in figure 1, allows it to be applied to a bifacial solar module using a transparent back sheet such as glass. When we install this module, scattered light and light that is reflected from the ground are incident on the solar module from the backside. In our experiments, the output power of a bifacial solar module was higher by around 20% than that of a standard HIT solar module. That value depends on mounting conditions, such as the mounting angle, the ground reflectance, and the interval between modules.

Vertical mounting allows us to apply it, for example, to a fence, a door, or a handrail. Horizontal mounting allows us to apply it, as shown in figure 9, to the roof of a bus stop.

8

Figure 9. HIT Double used for the roof of a bus stop in Osaka, Japan.

Figure 10 shows output power by dependence on the mounting tilt angle. The output power of the bifacial modules is higher than those of the standard module. We effectively increased the output power of the bifacial solar modules by increasing the tilt angle over that of standard modules. At a tilt angle of 60 degrees, the increase of output power relative to that of the standard module is 25%. Meanwhile, the relative increase is 17% at a tilt angle of 30 degrees. This allows the HIT Double to be used in a wide range of applications. It should, however, be noted that the amount of increase of output power with bifacial solar modules greatly depends on the installation environment.

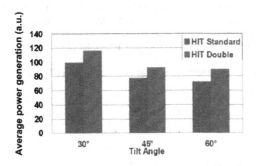

Figure 10. A comparison of output power between standard and bifacial modules with various mounting tilt angles. Values are normalized at that of a HIT standard module with a tilt angle of 30 degrees. Ground reflectance: 0.4, Module height from the ground: 3.28 feet, Facing direction: south. Date: May 22 - June 21.

SUMMARY

We have described the excellent power generating properties of HIT cells.

We have been focusing on the following techniques for obtaining high conversion efficiency with the HIT structure. We plan to raise the cell conversion efficiency to more than 23.5% in the laboratory by the end of 2010. To do this, we must simultaneously improve the HIT structure to increase the V_{oc}, optimize the grid electrode and reduce the absorption loss to increase the I_{sc}.

In terms of utilizing thinner Si wafers, we have exhibited a high potential for use on thinner c-Si wafers. We obtained a high conversion efficiency of 21.4% and an extremely high V_{oc} value of 0.739 V (AM 1.5, 1 SUN, 25°C, cell thickness: 85 μm, cell size (AP): 103.3 cm^2 measured in-house). The V_{oc} increases as the thickness of the c-Si wafer decreases, due to the extremely low surface recombination velocity with excellent passivation on the c-Si surface.

The excellent temperature coefficient results in generating more output power at high temperatures by further improving the V_{oc}, such as by thinning the wafer and improving the HIT structure.

The application of the HIT solar cell to a bifacial solar module has been commercialized as the "HIT Double," which is used for a wide range of practical applications for generating electricity.

Consequently, the excellent power generating properties of HIT solar cells will make it possible to reduce the power generating cost.

REFERENCES

1. S. Taira et al., "Our Approaches for Achieving HIT Solar Cells With More Than 23% Efficiency", 22nd EU-PVSEC, pp. 932-935 (2007).
2. H. Kanno et al., "Over 22% Efficient HIT Solar Cell", 23rd EU-PVSEC (2008)(in press).
3. M. Taguchi et al., "Improvement of the Conversion Efficiency of Polycrystalline Silicon Thin Film Solar Cell", Fifth PVSEC, pp. 689-692 (1990).
4. M. Tanaka et al., "Development of New a-Si/c-Si Heterojunction Solar Cells: ACJ-HIT (Artificially Constructed Junction-Heterojunction with Intrinsic Thin-Layer)", Jpn. J. Appl. Phys. 31, pp. 3518-3522 (1992).
5. Y. Tsunomura et al., "Twenty-two percent efficiency HIT solar cell", Solar Energy Materials & Solar Cells (2008).
6. D. Ide, et al., "Excellent power-generating properties by using the HIT structure", 33rd IEEE PVSEC (2008).
7. R. Brendel, "SUNRAYS: A Versatile Solar Cell Ray Tracing Program for the Photovoltaic Community", Twelfth EC Photovoltaic and Solar Energy Conf., p. 1339 (1994).
8. S. Taira et al., "Temperature Properties of High-Voc HIT Solar Cells", Renewable energy, pp. 115-118 (2006).
9. A. Terakawa et al., "High Efficiency HIT Solar Cells and the Effects of Open Circuit Voltage on Temperature Coefficients", Fifteenth PVSEC (2005).

Mater. Res. Soc. Symp. Proc. Vol. 1123 © 2009 Materials Research Society   1123-P03-06

Effect of thermal annealing on characteristics of polycrystalline silicon used for solar cells

Xianfang GOU[1], Xudong LI [1], Ying XU[1], Xinqing Liang[2], Yuwen ZHAO[1]
1. Beijing Solar Energy Research Institute, HaiDian District, HuaYuanLu3#, Beijing, china
2. Department of Physics, UC Santa Cruz, 1156 High Street Santa Cruz, CA 95064

ABSTRACT

Behavior of oxygen, carbon and minority-carrier lifetimes of multi-crystalline silicon (mc-Si) has been investigated by means of FTIR and QSSPCD after three step annealing. For comparison, the annealing of czochralski (CZ) silicon was also carried out under the same conditions. The results revealed that oxygen and carbon concentration of mc-Si had a larger decrease than that of CZ -Si, which means the more oxygen precipitates in mc-Si were generated. High density defects of mc-Si such as grain boundaries and dislocations accelerated formation of oxygen precipitates. Bulk lifetime of mc-Si and CZ -Si greatly increased. The reason of lifetime increase is probably due to the fact that lots of oxygen precipitates and defect complex compounds are generated after three step annealing, which could be suction center of defect, that reduced carrier decentralized recombination centers that resulted in improvement of lifetime of wafer. Tendency of difference of lifetime was correlated with interior structure of crystalline silicon.

INTRODUCTION

Polycrystalline silicon is now base material for production of photovoltaic converters of a solar energy. However, they need further quality improvement for highly efficient and low-cost solar cells by understanding the behavior of impurities and defects in the polycrystalline Si wafers in more detail. Because there are grain boundaries and more impurities and defects, mc-Si material has more complicated physical behavior in high temperature annealing. Oxygen in mc-Si is a very important impurity that has an influence on the electrical and mechanical properties of silicon material during heat treatments. The paper investigated three-step annealing effect on lifetimes and oxygen concentrations.

EXPERIMENTAL

In this experiment, wafers were p-type ($0.75\sim1.2\Omega \cdot$ cm), thickness of 290μm, provided by Bayer Solar Corporation. For comparison, CZ-Si samples, p-type($1\sim3\Omega\cdot$cm), <100> orientation, thickness of 330μm were also studied. The samples were cleaned with chemical solution, and silicon oxide was removed in HF(10%) solution. The samples were put in an annealing

Corresponding author: +86-01-62001053   Fax: +86-01-62001022
Emai address: gou_xian_fang@163.com

furnace in 1260°C for one hour, N_2 acting as protection gas so that getting rid off effect of thermal history[1]. Then, the experiments were put in an annealing furnace in 1150°C(2h)+750°C(6h)+1000°C(4h). Finally, the samples were etched by Wright solution and defects were observed by optical microscope(OLYMPUS, STM6) and SEM.

Fourier transmission infrared spectroscopy (FTIR) was used to measure the concentrations of interstitial oxygen. Minority carrier lifetime was surveyed by means of quasi-steady state photoconductance (QSSPCD) after annealing.

DISCUSSION

Effect of thermal annealing on oxygen concentration

As shown in Table 1. CZ silicon and mc-Si samples, oxygen concentrations decreased obviously after three step annealing in N_2 ambient, Oxygen concentrations of mc-Si samples have more decrease, which means high density defects such as grain boundary and dislocation that accelerated generation. Oxygen precipitates formation were contact with initial interstitial oxygen concentrations. Some reacher reported[2]that impurity carbon formatted microdefects with volume contraction effect that as nucleating center. Specially, impurity carbon might heavy accelerated oxygen precipitates formation in lower interstitial oxygen concentrations silicon. Reason of carbon concentration decrease possible that carbon as nucleating center oxygen precipitate accelerated oxygen precipitates formation in thermal annealing.

Table1. Concentrations of oxygen and carbon of samples after three step annealing

samples	$[O_i]\times10^{17}$atms/cm^3		$[C_s]\times10^{17}$atms/cm^3	
	before	after	before	after
Cz-1	8.3	7.9	0.22	0.19
mc-1	8.5	5.7	2.97	2.25
mc-2	6.2	4.4	2.21	1.9
mc-3	5.8	2.7	2.18	1.16

During three step annealing process, the first step annealing show that some diffused to silicon surface at higher temperature. The second step annealing at 750°C show that some interstitial oxygen atom accumulated to be nucleating center. The third step, annealing at 1000°C, show that oxyge precipitate gradually began to grow. Oxygen precipitate induced defect with growth. Because of volume change created oxygen precipitate, which resulted to generate stress field around oxygen precipitate. But oxygen precipitate decreased stress by emitting self-interurestitial oxygen atom, so which formed the non-intrinsic dislocation. As shown in Figure1. and Figure 2., Defect photograph of mc-1 sample optical micrograph (500×) and SEM. There were a large number of triangle dislocation of mc-Si.

12

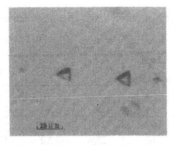

Figure1. Defect photograph of mc-1 sample optical micrograph (500 ×)

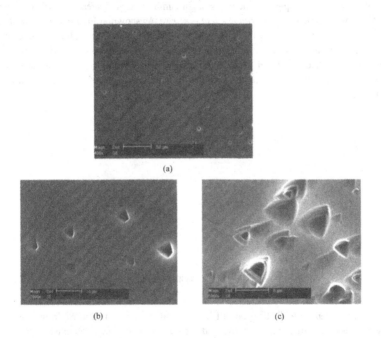

(a)

(b)   (c)

Figure 2. SEM defect photograph of mc-1 sample after three step annealing
(a)400 ×   (b)2000 ×   (c)5000 ×

Effect of thermal annealing on carrier lifetime

Table2. Changes of carrier lifetime of silicon wafers after three step annealing

samples	Lifetime(μs)			
	before annealing	1150℃	750℃	1000℃
Cz-1	2.2	0.2	5.5	17.9
mc-1	14	20	12.4	21.2
mc-2	14.5	24.4	10.6	26
mc-3	12	22.4	12.2	23.9

Table2 gives changes of carrier lifetime of silicon wafers after three step annealing. After annealing the Cz-1 sample lifetime was straight climb but light decrease at 1150℃. Lifetime of CZ-Si showed tendency of *decrease – rise – re-rise*. As shown in Figure 4. document reported that new recombination center that many oxygen precipitate and secondary defects generated in CZ-Si at first step annealing. so carrier lifetime directly fall down. During later two step annealing, the defects formed complex compound that induced some stress, which reduced decentralized recombination center because of gettering of electrically active impurities, so lifetime increased.

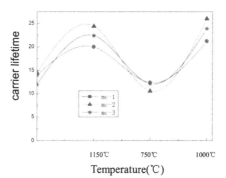

Temperature(℃)

Figure 3. Changes of carrier lifetime of silicon wafers after three step annealing

As shown in Figure3, tendency of lifetime curve of mc-Si differed from that of CZ-Si. Lifetime tendency of mc-Si was rise – decrease – re-rise. The phenomenon possibility that because a large number of oxygen atoms in mc-Si diffused to sample surface and escaped surface at high temperature. Other reason probably that primary defects decomposed and resulted to recombination center decrease, so lifetime increased. In the second step annealing, some interstitial oxygen atom accumulated to becoming nucleating centers and body defect cha impelled lifetime to fall. The third step annealing, the defects formed complex compound as getter center to improve lifetime.

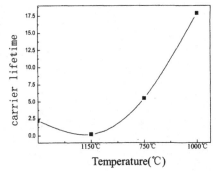

Temperature(℃)

Figure 4. Changes of carrier lifetime of silicon wafers after three step annealing

CONCLUSION

The results revealed that oxygen and carbon concentration of mc-Si had a greater decrease than that of CZ -Si, which means the more oxygen precipitates in mc-Si were generated. High density defects of mc-Si such as grain boundaries , dislocations accelerated formation of oxygen precipitates. Bulk lifetime of mc-Si and CZ–Si greatly increased. The reason of lifetime increase is probably considered due to the fact that lots of oxygen precipitates and defect complex compound generated after three step annealing, which could be suction center of defect. That reduced carrier decentralized recombination centers that resulted in improvement of lifetime of wafer. Tendency difference of lifetime was correlated with interior structure of crystalline silicon. Effect of three step annealing on mc-Si solar cell needs further research.

ACKNOWLEDGMENT

The authors would like to thank following finance support. *This project was supported by BeiJing bud plan* and *the National Natural Science Foundation* （60576065）and *National 863 project* (2007AA05Z437).

REFERENCES

1 D. Yang, Dongsheng Li . "Oxygen in Czochralski silicon used for solar cells " , Solar Energy Materials & Solar cells.   72.133 (2002)
2 Rivand L. Anognostopoulos C N, Erikson G R,J. Electrochem,Soc.1988. P135-437.

15

Mater. Res. Soc. Symp. Proc. Vol. 1123 © 2009 Materials Research Society   1123-P03-09

Conducting two-phase silicon oxide layers for thin-film silicon solar cells

Peter Buehlmann[1], Julien Bailat[2], Andrea Feltrin[1], Christophe Ballif[1]

[1]IMT, University of Neuchâtel, Neuchâtel, Switzerland
[2] Now at Oerlikon Solar-Lab, Neuchâtel, Switzerland

ABSTRACT

We present optical properties and microstructure analyses of hydrogenated silicon sub-oxide layers containing silicon nanocrystals (nc-SiO$_x$:H). This material is especially adapted for the use as intermediate reflecting layer (IRL) in micromorph silicon tandem cells due to its low refractive index and relatively high transverse conductivity. The nc-SiO$_x$:H is deposited by very high frequency plasma enhanced chemical vapor deposition from a SiH$_4$/CO$_2$/H$_2$/PH$_3$ gas mixture. We show the influence of H$_2$/SiH$_4$ and CO$_2$/SiH$_4$ gas ratios on the layer properties as well as on the micromorph cell when the nc-SiO$_x$:H is used as IRL. The lowest refractive index achieved in a working micromoph cell is 1.71 and the highest initial micromoph efficiency with such an IRL is 13.3 %.

INTRODUCTION

Thin film silicon technologies have a high potential for further cost reduction of terrestrial photovoltaics for energy production. In this technology, multi-junction solar cells are candidate to achieve efficiencies comparable to the well established crystalline silicon solar cells. One of the most promising structures is the micromorph cell where a high band-gap amorphous top cell and a low band-gap microcrystalline bottom cell are stacked upon each other to better exploit the solar spectrum. The individual cells composing a multi-junction cell are electrically connected in series and special care has therefore to be taken to match their currents. Matching of currents can be achieved by adjusting the thickness of the thin absorber layers or by the insertion of an intermediate reflecting layer (IRL). The second solution is especially interesting for the micromorph cell where the top cell has to be kept reasonably thin (< 300 nm) to limit the effects of the light induced Staebler-Wronski degradation of the amorphous material [1]. For high reflectivity, the IRL should have a refractive index (n) as low as possible whereas for the electrical requirements, the IRL must have a transverse conductivity of at least 10^{-5} S/cm to avoid blocking the device. Very low in-plane conductivity is desired to avoid series interconnection of shunts in the elementary cells. The first results on micromorph cells with IRL were presented in 1996 by D. Fischer [2] who used ZnO as low-n material. Later a different approach was introduced by K. Yamamoto [3] without specifying the material used. In this study we will investigate on the use of n-doped, hydrogenated silicon sub-oxide containing silicon nanocrystals (nc-SiO$_x$:H) as proposed by our group [4] and A. Lambertz [5] in 2007.

EXPERIMENTAL DETAILS

The n doped nc-SiO$_x$:H films were prepared from a mixture of SiH$_4$, CO$_2$, H$_2$ and PH$_3$ gases in a capacitively coupled very high frequency plasma enhanced chemical vapor deposition (VHF-PECVD) system. In this study all the SiO$_x$ layers were deposited at a pressure of 0.7 mbar, at a frequency of 110 MHz, with a power density of 0.1 W/cm^2, a constant PH_3/SiH_4 gas ratio of 0.1 and a deposition time sufficient to reach a layer thickness of ~ 90 nm. Ellipsometry, Raman scattering and thickness measurements were performed on layers deposited on glass. Refractive index (n) was found from fitting the ellipsometry measurements to a Tauc-Lorentz dispersion model including a surface roughness layer. The thickness was measured with a height profiler. Infrared (IR) absorption measurements were performed with a Nicolet 8700 system from Thermo on samples deposited on intrinsic, one-side polished wafers. The absorption spectra were normalized with the layer thickness.

The amorphous and microcrystalline cells were prepared by PECVD. Glass covered with rough CVD ZnO was used as substrate for the micromorph cells. CVD ZnO was used as back contact and teflon as diffusive back reflector. The cells were fully patterned to a size of 1.2 cm^2. The short-circuit current density (J_{sc}) of the solar cells was calculated from the measurement of the external quantum efficiency (EQE) curve, by integrating, over the wavelength range from 350 to 1100 nm, the product of EQE times the incoming photon flux of the AM1.5g solar spectrum. The current–voltage curves were measured under a dual lamp WACOM solar simulator in standard test conditions (25 °C, AM1.5g spectrum, and 1000 W/m^2).

RESULTS

In this study, we present nc-SiO$_x$ layers from three different gas ratio series. In the series H100 and H300, we varied the CO_2/SiH_4 between 1 and 3.5 with a constant H_2/SiH_4 gas flow ratio of 100 and 300 respectively. In series C2.5, the H_2/SiH_4 was varied from 100 to 400 with a constant CO_2/SiH_4 gas ratio of 2.5. Figure 1 compares the growth rates of all the nc-SiO$_x$ layers from these three series.

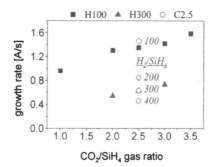

Figure 1. Growth rates of nc-SiO$_x$ layers from the three series: H100, H300 and C2.5

Figure 2a shows the IR absorption spectra of the H100 series. The peaks at ~ 1050 cm^{-1} and between 800 and 900 cm^{-1} which have been reported to stem from the Si-O-Si stretching and bending modes [6, 7] increase with increasing CO_2/SiH_4 gas ratio. The oxygen content increases in a similar manner when the H_2/SiH_4 is increased from 100 to 300 with a fixed CO_2/SiH_4 of 2.5 as it can be seen on figure 2b.

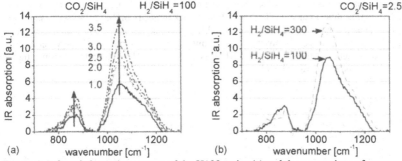

Figure 2. Infrared absorption spectra of the H100 series (a) and the comparison of two spectra with CO_2/SiH_4 gas ratio of 2.5 but different H_2/SiH_4 gas ratios (b).

Figure 3a and 3b show the refractive index of the nc-SiO$_x$ layers as a function of CO_2/SiH_4 and H_2/SiH_4 gas ratios. For the series H100 and H300, the n decreases with increasing CO_2/SiH_4 gas ratio. The same effect is also observed for the C2.5 series with increasing H_2/SiH_4 gas ratio. This tendency has been described by Iftiquar [6] who suggested that the main mechanism of oxygen incorporation in SiO$_x$ films deposited by PECVD is through the formation of O-H complexes in the plasma which more easily get to the growing layer surface than the highly reactive and electronegative atomic oxygen.

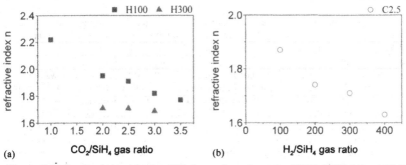

Figure 3. Refractive index (n) of nc-SiO$_x$ layers from the series: H100, H300 (a) and C2.5 (b)

To gain further insight into structural properties of the nc-SiO$_x$ layers we performed Raman scattering measurements and calculated the Raman crystalline fraction in the silicon phase [8]. Since the measurements don't account for the SiO, which is the dominant phase in our layers, the signal represents in this case not a crystalline volume fraction of the layer, but only the ratio of crystalline silicon signal over the crystalline plus amorphous silicon signal. The Raman scattering signal is very noisy because there is only little silicon phase remaining in our films. Figure 4a shows the Raman spectra of the H100 series. Figure 4b and 4c show the calculated Raman crystalline fraction in the silicon phase, and the total Raman silicon signal intensity as a function of CO_2/SiH_4 gas ratio for the H100 and H300 series. The crystalline fraction decreases with increasing CO_2/SiH_4 gas ratio whereas the total silicon Raman signal intensity remains almost constant. The total Raman peak intensity of the H100 series is almost twice as high as for the series H300 for comparable crystalline fractions. Based on these data, increasing CO_2/SiH_4 seems to convert the crystalline silicon to amorphous silicon without reducing the presence of the silicon phase whereas an increased H_2/SiH_4 seems to reduce the presence of silicon phase significantly.

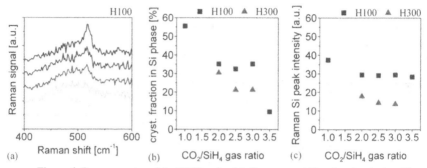

Figure 4. Raman spectra of the H100 series (a), Raman crystalline fraction in the silicon phase (b) and total Raman silicon peak intensity (c) of the H100, H300 and C2.5 series.

DISCUSSION

When the nc-SiO_x layers are used as intermediate reflectors in micromorph cells, a sufficiently high transverse conductivity is needed to avoid blocking the device. If a 100 nm thick layer should not increase the series resistance in the micromorph cell more than 1 Ohm·cm^2, we can calculate a lower limit of the conductivity of 10^{-5} S/cm. On all the layers presented above we measured in-plane conductivities below this value. Figure 5 shows the current density-voltage (JV) curves of top-limited micromorph cells with the nc-SiO_x layers from the H100 and the H300 series. J_{sc} increases with decreasing n of the silicon oxide based intermediate reflector (SOIR) and there are cells with no increase of series resistance due to the SOIR with n as low as 1.71 although the measured in-plane conductivity of most SiO_x layers is below 10^{-10} S/cm, which is the lower limit of our measurement system. From figure 4 we can assume that the in-plane conductivity is mostly determined by the highly resistive amorphous SiO_x matrix whereas we belief that silicon nanocrystals are responsible for a relatively high transverse conductivity as observed in the micromorph cells.

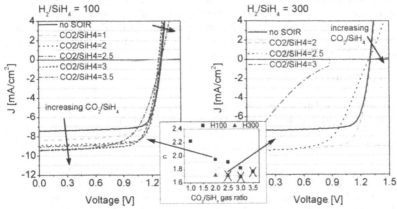

Figure 5. J-V curves of top-limited micromorph solar cells with the different SOIR from the H100 and H300 series. The inset shows the refractive index of the SOIR. SOIR that caused an increase of series resistance in cell are marked with a cross).

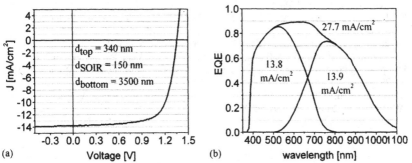

Figure 6. J-V (a) and EQE (b) curves of micromorph cell with 13.3 % initial efficiency (Cell fully patterned to 1.2 cm^2; with broadband anti-reflection coating on glass; V_{oc} = 1.36 V, FF = 70.8, J_{sc} = 13.8 mA/cm^2)

The SOIR with the lowest refractive index (n = 1.71) without deteriorating the series resistance of the micromoph cell is deposited with CO_2/SiH_4 = 2 and H_2/SiH_4 = 300. Figure 6 shows an optimized micromorph cell, where the use of a SOIR has lead to a best initial efficiency of 13.3 % with top cell, SOIR, bottom cell thicknesses of 340, 150, 3500 nm respectively [9].

CONCLUSIONS

We presented in this study nc-SiO$_x$ layers with refractive indexes as low as 1.71 that can be used as SOIR in micromorph cells without compromising its series resistance. From optical measurements of the layers and from the results of the same layers incorporated as intermediate reflectors in micromorph cells we conclude that the high H_2 dilution is important to get low refractive index material with sufficiently high transversal conductivity. Finally we showed an optimized micromorph cell with an initial efficiency of 13.3 % with top cell, SOIR, bottom cell thicknesses of 340, 150, 3500 nm respectively.

ACKNOWLEDGMENTS

The authors gratefully acknowledge the support of the Swiss Federal Energy Office (OFEN) (project 101191) and the EU (Athlet Project, contract 019670).

REFERENCES

1. M. S. Bennett, J. L. Newton, and K. Rajan, Proceedings of the 7th European Photovoltaic Solar Energy Conference, Sevilla, Spain (Reidel, Dordrecht, 1987), pp. 544-548.
2. D. Fischer et al. in proceedings of the 25th IEEE Photovoltaic Specialists Conference, Washington D. C., USA (IEEE, New York, 1996), pp. 1053-1056.
3. K. Yamamoto et al., Solar Energy **77** (2004) pp. 939–949.
4. P. Buehlmann et al., Appl. Phys. Lett. 91, 143505 (2007).
5. A. Lambertz et al., 22th European Photovoltaic Solar Energy Conference, Milan, Italy (WIP, Munich, 2007), pp. 1839-1842.
6. S. M. Iftiquar, J. Phys. D: Appl. Phys. **31**, 1630 (1998); High Temperature Material Processes, **6** (2002)
7. G. Lucovsky et al., Phys. Rev. B **28**, 3225 (1983); Sol. En. Mat. **8,** 165 (1982).
8. C. Droz, E. Vallat-Sauvain, J. Bailat, L. Feitknecht, and A. Shah, Proceedings of 17th European Photovoltaic Solar Energy Conference, Munich, Germany (WIP, Munich, 2002), pp. 2917-2920.
9. D. Dominé et al., 23th European Photovoltaic Solar Energy Conference, Valencia, Spain (WIP, Munich, 2008), pp. 2091-2095.

Mater. Res. Soc. Symp. Proc. Vol. 1123 © 2009 Materials Research Society   1123-P05-24

METHOD OF FAST HYDROGEN PASSIVATION TO SOLAR CELL MADE OF CRYSTALLINE SILICON

Wen-Ching Sun[1], Jian-Hong Lin[1], Wei-Lun Chang[1], Tien-Heng Huang [2],Chih-Wei Wang[3], Jia-De Lin[3], Chwung-Shan Kou[3], Jian-You Lin[4], Sheng-Wei Chen[4], Jenn-Chang Hwang[4], Jon-Yiew Gan[4]

[1]Photovoltaics Technology Center Industrial, Technology Research Institute, Hsinchu, Taiwan
[2]Material and Chemical Research Laboratories Intranet, Technology Research Institute, Hsinchu, Taiwan
[3]Department of Physics, National Tsing Hua University, Hsinchu, Taiwan
[4]Department of Materials Science and Engineering, National Tsing Hua University, Hsinchu, Taiwan

ABSTRACTS

Plasma immersion ion implantation (PIII) is a technique of material processing and surface modification, using controllable negative high voltage pulsed bias to attract the ion generated from the plasma. The method using PIII treatment quickly improves the performance of solar cell made of crystalline silicon, including monocrystalline, multicrystalline and polycrystalline silicon. Hydrogen ions are attracted and quickly implanted into solar cell under a predetermined negative pulse voltage, thus, the passivation of the crystal defects of the solar cell can be realized in a short period. Meanwhile, the properties of the antireflection layer can not be damaged as the proper operating conditions are used. Consequently, the series resistance can be significantly reduced and the filling factor increases as a result. Both the short-circuit and the open-circuit voltage can be increased. The efficiency can be enhanced.

INTRODUCTION

Solar cell is a very promising clean energy source which can generate electricity directly from sunlight. However, the cost of the production of solar cells needs to be significantly reduced so as to be widely accepted as a major electricity source. It has been pointed out that the silicon wafer share is above one third of the total cost of a crystalline silicon solar cell module. Consequently, there are intensive researches on the development of solar cells based on multicrystalline silicon (mc-Si) or polycrystalline silicon. On the other hand, both mc-Si and poly-Si contain defects within the crystals, including grain boundaries, intragrain dislocations and precipitates. Those imperfections can degrade the conversion efficiency of solar cells. Besides, the recombination of charge carriers at the surface is detrimental to solar cells, even in the case of monocrystalline solar cells.

Dangling bonds on the surface are the main trapping centers for the charge carriers. Impurities can be removed by gettering and defects can be passivated[1] . Hydrogen atoms are believed to play a vital role in the deactivation of recombination centers. As a result, the efficiency of c-Si solar cells can be significantly improved. The general view has

been that these efficiency improvements are closely related to the reduction of the minority carrier recombination losses at surfaces, grain boundaries, dislocations and other defects in the crystal lattice. It is generally accepted that hydrogen is associated in a number of phenomena including dopant deactivation, passivation of deep levels due to defects and formation of other complexes. The capability of hydrogen to passivate defects liked is locations and grain boundaries, has been quite success fully applied for improving performance of lower grade polycrystalline silicon solar cells. There is, hence, interest in understanding behavior of hydrogen in Si from general as well as technological point of view. Typically, solar cells are hydrogenated from the junction side using an rf plasma or low energy ion implantation which can result in high surface concentration of hydrogen[2] . It has, however, been suggested that under high surface concentration, hydrogen can acquire any of the following forms interstitial hydrogen molecule, a plane of Si-H bonds or a plane of hydrogens occupying bond-center sites[3] . There are many successfully established means of hydrogen passivation such as hydrogen ion implantation, hydrogen plasma injection, plasma enhanced chemical vapor deposition (PECVD) of hydrogenated silicon nitride, and forming gas annealing (FGA).

Plasma Immersion Ion Implantation (PIII) was developed for the beneficial modification of surface-sensitive properties[4] . In analogy to conventional beam-line ion implantation, it uses energetic ions, mostly nitrogen, that are implanted into near-surface region of material. In this paper, silicon crystalline solar cell is immersed in hydrogen plasma and subjected to negative high-voltage pulses. In the electrical field, the hydrogen ions are accelerated to high energies and incorporated into the solar cell.

EXPERIMENTAL DETAILS

Figure 1 shows cross-sectional front view of a typical solar cell. and Figure 2 shows the corresponding processing flow sequence.

Initially, we used (100)-oriented Czochralski (Cz) and multi-crystalline boron-doped p-type Si wafers of 1.5Ωcm resistivity and 200μm thickness. The wafers were cleaned in a dilute HF solution prior to solar cell standard processing to remove the native oxide. Then, thermal oxide was grown using a wet oxidation furnace after texturization. Using a standard process for acidic etch recipe with texturization can remove saw damage in one single process step.

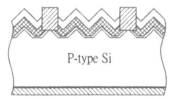

Figure 1. is a cross-sectional front view of a typical solar cell.

Later, gaseous oxygen was passed through the heated DI water and led to the furnace. The temperature of furnace was below 850°C. The thickness of the thermal oxide was around 100nm. The oxide on the front side was removed to allow subsequent diffusion. The emitter was formed on the front surface by diffusion using POCl₃ liquid as a doping source. The diffusion temperature was 840°C, and the flow rate of POCl₃ was 600sccm. The sheet resistance of the n^+ emitter was around 50 Ω / square emitter.

After the diffusion processes, all wafers received a standard cleaning followed by etching with a diluted hydrofluoric acid (HF) immediately before growing passivating layers. Then, the wafers were operated in RCA process after forming gas at 500°C for 30mins, resulting in a ~1.5 nm thick oxide layer. After rinsing with de-ionized water, Then, the SiNx:H was deposited on front side of these wafers followed by plasma-enhanced chemical vapor deposition (PECVD) for antireflection purposes. The thickness of the antireflection layer was 90nm, and its reflective index was approximately 1.9. After the antireflection coating was deposited, the front and back electrodes (Ag paste: Du Pont PV145 and Al paste: Du Pont PV333) was printed and put in an IR furnace at the maximum zone temperature of 965°C. To compare the solar cell devices, all the groups were put in IR heated belt furnace under the same conditions (see Figure 2.).

After solar cell fabricated process, the PIII was applied. The base pressure of the vacuum chamber is 10^{-6} Torr, and then hydrogen gas is intruded into the vacuum chamber as a working gas and the pressure is raised to 2 mTorr. The plasma is excited by a RF power (13.56 MHz) through an inductive coupling antenna with a power of 200 W. The plasma density is approximately 10^{11} cm⁻³. Furthermore, a bias is applied to the solar cell by a pulse voltage of -4 kV. The pulse width is 10 μsec and the pulse frequency is 200 Hz. In this experiment, no power supply is provided to heat the solar cell, but the temperatures of the samples are approximately 100°C resulting from the plasma ions implantation. The total process time is 10 min.

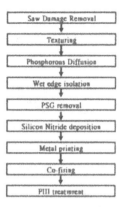

Figure 2. Process sequence of the fabrication of the sample solar cell.

Results and Discussion

Independently varied ion implantation parameters include ion energy, dose-rate and implantation time. Defects in the silicon are expected to be passivated by deeply diffused hydrogen, which can be enhanced by higher dose rate, higher sample temperature and lower implantation time. Implantation damage or passivation were also observed with increasing energy, sample temperature and implantation time.

Type	Treatment	Isc (A)	Voc (V)	FF (%)	Eff (%)
Cz cell	Before PIII	0.23	0.593	75.03	14.25
	After PIII	0.25	0.602	80.77	17.06
Multi cell	Before PIII	0.195	0.586	76.99	12.33
	After PIII	0.199	0.591	81.25	13.39

Table I: Comparison of conversion efficiency of cells, before and after hydrogen passivation under WACOM solar simulator (AM1.5G, 100 mW/cm², 25 °C).

Table I shows the results of the solar cell parameters using before and after hydrogen passivation measured by a class-A solar simulators. Figure 3 is the comparison of current-voltage characteristics the solar cell before and after hydrogen passivation. It is shown by the results that the series resistance is significantly reduced and the fill factor(FF) increases from 76.99% to 81.25%, and the short-circuit current is increased. These improvements lead to an increase of the conversion efficiency(Eff.) from 12.33% to 13.39 %. And, Figure 4 is the comparison of current-voltage characteristics the solar cell before and after hydrogen passivation. It is shown by the results that the fill factor increases from 75% to 80.77 % as a result. Meanwhile, the short-circuit current(Jsc) increases from 0.23 A to 0.25 A and the open voltage(Voc) increase from 0.591 V to 0.602 V as well. These improvements lead to an increase of the

24

conversion efficiency from 14.25 % to 17.06 %.

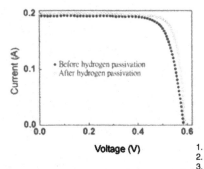

Voltage (V)

Figure 3. is a plot of the electrical characteristics (I-V) of a solar cell made of a multicrystalline silicon illustrated in FIG. 1 before and after hydrogen passivation, under simulated AM1.5 illumination

The improvement in fill factor can be explained in this case by decreases in contact resistance of front side electrode. This was verified comparing with cells exposed to similar hydrogenation conditions.

The improvement in short circuit current density and open circuit voltage was related to hydrogen passivation. For the passivation of defects, impurities and segregated impurities on extended defects, hydrogen can play an important role.

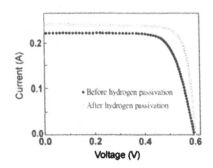

Voltage (V)

Figure 4. is a plot of the electrical characteristics (I-V) of a solar cell made of monocrystalline silicon as illustrated in FIG. 1 before and after hydrogen passivation, under simulated AM1.5 illumination.

Conclusion

In view of above, compared with the existing techniques, the PIII treatment can significantly reduce the time and the cost of hydrogen passivation, and effectively improve the efficiency of c-Si solar cells. Furthermore, the implements of this method are simpler and more economical in a mass production process. It can be used as an easy step for passivation of defects and impurities of commercial silicon solar cell. Especially, it can perform hydrogen passivation to the solar cells which fails to meet the requirements for efficiency in the production, so as to improve the efficiency and increase the production yield. In addition, the PIII treatment is not required to change the existing production methods of solar cells, so it is independent process and has high conformability

References

1. Yagi H, Matsukuma K, Kokunai S, Kida Y, Kawakami N, Nishinoiri K, Saitoh T, Shimokawa R, Morita K. Proceedings of the 20th IEEE PVSC, vol. 2. 1988, p. 1600.

2. J. I. Hanoka e t . a l , Appl. Phys. Lett., 42, 618 (1983).

3. S. J. Feng, e t . a l , Appl. Phys. Lett., 53,1735 (1988).

4. Seager, C. H., Sharp, D. J., Panitz, J. K. G. and Aiello, R. V., J. Vac. Sci. Technol. 20 (1985) 430.

5. J. Zhao, A. Wang and M.A. Green. Prog. Photovoltaics 7, 471 (1999).

6. A. W. Stephens, A. G. Aberle, M. A. Green, J. Appl. Phys. 76 , 363 (1993).

7. S. K. Dhungel, J. Yoo, K. Kim, B.Karunagaran, Materials Science in Semiconductor Processing 7, 427–431 (2004)

8. J. C. Muller, A. Barhdadi, Y. Ababou and P. Siffert, Revue Phys. Appl. 22, 649-654 (1987)

9. H.F.W. Dekkers! , S. De Wolf, G. Agostinelli, F. Duerinckx, G. Beaucarne. Sol. Energy Mater. 90, 3244–3250 (2006)

10. P. Sana, A. Rohatgi, J. P. Kalejs, and R. O. Bell, Appl. Phys. Lett. 64, 97 (1994).

11. W. Schmidt, K.D. Rasch, and K. Roy, 16 IEEE Photovoltaic Specialist Conference, San Diego, 1982, pages 537-542,.

12. R. Hezel and R. Schroner, J. Appl. Phys., 52(4), 3076 (1981)

13. J. E. Johnson, J. I. Hano Ka, and J. A. Gregory, 18 IEEE Photovoltaic Specialists Conference, Las Vegas 1985, pages 1112-1115.

14. M. Spiegel, R. Tolle, C. Gerhards, C. Marckmann, N. Nussbaumer*,P. Fath, G. Willeke, E. Bucher, 26th IEEE Photovoltaic Specialists Conference, Anaheim, 1997, pages 151-154.

Mater. Res. Soc. Symp. Proc. Vol. 1123 © 2009 Materials Research Society 1123-P02-08

Effects of Deposition Parameters on the Structure and Photovoltaic Performance of Si Thin Films by Metal Induced Growth

Peter T. Mersich[1], Shubhranshu Verma[1], Wayne A. Anderson[1], and Rossman F. Giese Jr.[2]
[1]Department of Electrical Engineering, University at Buffalo, State University of New York, Buffalo, NY 14260, U.S.A.
[2]Department of Geology, University at Buffalo, State University of New York, Buffalo, NY 14260, U.S.A.

ABSTRACT

A metal-induced growth (MIG) process was employed to deposit thin films of microcrystalline silicon (μc-Si) for solar cell applications. Due to different grain orientations of the crystals, the absorption coefficient of μc-Si is about 10 times higher than the absorption coefficient of single crystalline Si. The properties of the Si film were investigated resulting from variations in several parameters. A range of Ni and Co thicknesses were examined from 7.5 nm to 60 nm including combinations of the two, while the dc sputtering power was stepped up from 150 W to 225 W. The structure of the resulting film was studied using scanning electron microscopy (SEM), energy dispersive x-ray spectroscopy (EDS) and x-ray diffraction (XRD). SEM of the film revealed that 5 hr of Si deposition at 150 W yields a film thickness of 6.5 μm and a maximum grain size of about 0.6 μm. EDS data showed that at the middle of the Si film the atomic percentage of the Si was 99.17%. XRD data showed that the dominant crystal orientation is {220}. To characterize the photovoltaic properties of the μc-Si, Schottky photodiodes were fabricated. Ni alone as the seed layer resulted in ohmic behavior. With Co only, MIG formed a rectifying contact with open-circuit voltage (V_{oc}). The combination of Co layered over Ni formed better thin films and gave a V_{oc} of 0.24 V and short-circuit current density (J_{sc}) of 5.0 mA/cm^2 since the Co prevents Ni contamination of the top of the grown Si layer.

INTRODUCTION

Amid the escalating energy crisis, there has been greatly expanded interest in establishing a viable, low-cost approach to solar cells [1]. While traditional wafer-based crystalline silicon solar cells are able to achieve relatively high efficiencies, this is easily outweighed by the cost of manufacturing. Therefore, many have turned to thin film solar cells [1]. Thin films not only have the advantage of much cheaper fabrication but also utilize far less material. This is made possible through the use of μc-Si, which has much higher optical absorption when compared to single crystal silicon due to different crystal orientations in the grains [2]. However, the effects of grain boundaries must then be taken into account, where larger grain sizes are desirable to minimize the adverse effects [3].

Metal-induced growth (MIG) is a method used to epitaxially grow thin layers of μc-Si [4]. Metal-induced crystallization (MIC) uses the two step process of depositing amorphous Si and later converting it to crystal form [5]. MIG has the advantage of a one step process to form the µc-Si and simultaneously form the back ohmic contact. When Si is sputtered from a target, a pre-deposited thin metal film produces a silicide seed layer that is used as a catalyst in the crystallization of Si at a temperature of 625°C. This seed layer, namely Ni silicide ($NiSi_2$) or Co silicide ($CoSi_2$), then has the advantage of a lattice mismatch of only 0.4% [6] and 1.2% [7], respectively, with Si as the sputtering process continues and the µc-Si film is grown. The size of the grains, atomic composition, and photovoltaic performance of the resulting films are all affected by various parameters during the deposition process. Several of these parameters were examined with respect to their effect on the µc-Si films. These parameters include the thickness of the metal layers and the power and time of sputtering. Also, the crystal structure of the film after the formation of the silicide layer and after the full completion of the process is compared.

EXPERIMENT

The MIG process began with a substrate of either oxide-coated silicon or tungsten, to demonstrate the feasibility of the process on a foreign substrate. An initial layer of Ni, Co, or a combination of the two was thermally evaporated on the substrate, which was then immediately taken for dc sputtering of Si. The thickness of these layers was varied from 7.5 nm to 60 nm. After the substrate was heated to 625°C, an n-type Si target of resistivity 0.2 Ω-cm was sputtered using a dc magnetron. A power of 50 W was first administered for 45 mins at which point the silicide seed layer was formed. The power was then increased to a higher level in order to establish the µc-Si film. This power level was varied from 150 W to 225 W. The time of sputtering was also adjusted from 3-5 hr to account for changes in the deposition rate. The resulting grains formed a columnar structure, which allowed for fewer grain boundaries through the vertical pathways of the film. A separate metallization was done to form the back contact for the silicon substrate, where the silicide layer was utilized. However, for the tungsten, the conductive metal substrate formed the back contact without any additional processing. The films were then annealed at 700°C for 2 hr in forming gas of 85% Ar and 15% H_2 to activate the dopants and relieve any stress in the film.

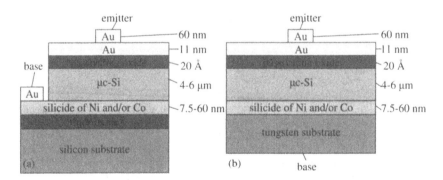

Figure 1. Diagram of Schottky photodiodes on (a) silicon and (b) tungsten substrates using the MIG process for the growth of µc-Si.

28

Schottky photodiodes were fabricated on the films in order to analyze the photovoltaic properties of the film. The film surface was cleaned with buffer HF. It was then placed in a furnace and ramped to 600°C to establish a thin native oxide, which passivated the surface. Au was then thermally evaporated to form the Schottky junction. A final diagram of the devices on each substrate is shown in Figure 1. Dark and 1.5AM photo measurements were carried out.

DISCUSSION

The μc-Si films were analyzed after the deposition. Structural analysis was conducted with the use of SEM, EDS, and XRD. Grain size and film thickness were confirmed with SEM. The atomic composition of the film was evaluated using EDS. In order to study the formation of crystal structures in the film, XRD was performed.

Structural Properties

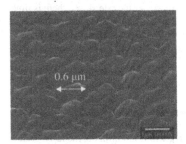

Figure 2. Large grain sizes achieved using Ni induced growth.

In order to study the effects of metal layer thicknesses on the structure of the film, the power and time of the deposition was held constant at 200 W for 3 hr. SEM revealed that regardless of metal used, a metal thickness of 7.5 nm did not result in proper grain formation. However, as the metal thickness was increased, the grain size increased. At a metal thickness of 60 nm, the grain size of 0.6 μm was the largest observed for Ni alone, as seen in Figure 2. For Co and Co/Ni, a similar effect was seen. However, the resulting grain sizes were substantially smaller with Co/Ni producing larger grains than Co alone.

29

 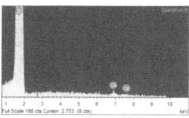

Figure 3. EDS of the μc-Si film at point 'A' revealed the atomic composition of Si to be 99.17%.

As the sputtering power was increased from 150 W to 225 W, the grain size decreased. This can be explained by the change in the rate of the deposition. As the rate slows, there is time for larger grains to be established. While larger grain size is advantageous, the reduced power caused the thickness of the film to suffer. Therefore, in order to achieve a similar thickness the sputtering was increased from 3 to 5 hr for the 150 W case. This allowed for an increase in thickness from 3 μm to 6.5 μm.

The atomic analysis using EDS was conducted at the middle section of films with comparable thickness using different metal layers. When Ni alone was used, there is a strong presence of Ni. This was due to the high diffusion constant of Ni in Si. However, when Co was used, the atomic percentage of Si was greater than 99% and reached 99.17% for films with Co/Ni, as shown in Figure 3. The high percentage of Si was maintained for the Co-coated Ni because Co was able to provide a diffusion barrier for the Ni, while still utilizing the large grain sizes achieved with Ni.

The analysis of XRD studied the transition of the silicide seed layer to the μc-Si film. The crystal structure was examined after the 50 W deposition in which the initial silicide seed layer is formed. This is compared to the data attained after the full 5 hr deposition. The results are shown in Figure 4. Figure 4a demonstrates a strong peak for NiSi {200} with smaller peaks of NiSi$_2$. However, as the deposition was fully carried out, the NiSi was converted, and NiSi$_2$ + Si {220} became the dominant peak, as seen in Figure 4b.

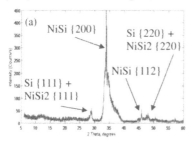 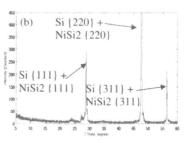

Figure 4. XRD showing the crystal structure (a) after the initial 45 mins of sputtering at 50 W and (b) after the full sputtering process.

Photovoltaic Properties

Photovoltaic measurements were conducted on the Schottky structures. In order to test the effects of metal thickness, MIG was performed at 200 W for 3 hr with different combinations of metal. These results are summarized in Table 1. When Ni alone was used, the current-voltage measurements revealed ohmic devices. This was due to the migration of Ni into the μc-Si as confirmed by EDS. With Co alone, rectifying diodes were achieved. However, the V_{oc} suffered from the small grains associated with Co induced growth. With the proper thickness of Ni and Co together, the V_{oc} was vastly improved, while attaining even higher J_{sc} values. This can be explained by increased grain sizes in addition to the prevention of Ni migration. The best performance was achieved at Ni and Co thicknesses of 23 nm and 45 nm, respectively.

Table I. J_{sc} and V_{oc} values obtained using different metal thicknesses

Ni thickness [nm]	Co thickness [nm]	J_{sc} [mA/cm^2]	V_{oc} [V]
7.5	-	-	-
60	-	-	-
-	7.5	-	-
-	30	2.8	0.052
-	60	0.6	0.02
25	5	0.5	0.01
25	25	1	0.075
25	45	2.5	0.19
23	45	3.6	0.205
45	45	0.9	0.13

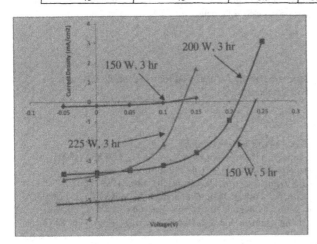

Figure 5. Changes in the photo current-voltage measurements resulting from changes in sputtering power and time.

For the analysis of sputtering power and time, the metal layers were maintained at the previously-obtained, optimal thicknesses. Figure 5 details the results obtained from these changes. As the power decreased from 225 W to 200 W, the V_{oc} improved due to the increase in grain size. However, when the power decreased to 150 W for the same sputtering time, the V_{oc} dropped significantly. From analysis of the μc-Si film through SEM, it was observed that the thickness was much smaller for the film at 150 W due to a decrease in the sputtering rate. Therefore, to obtain a comparable thickness, the sputtering time was increased to 5 hr. The resulting improvement in V_{oc} can again be associated with larger grain sizes.

CONCLUSIONS

The MIG process can be used to epitaxially deposit thin films of μc-Si. By manipulating the parameters involved, the structure can be modified to make it more desirable for solar cell application. By combining Co and Ni and increasing their thicknesses accordingly, the grain size can be increased, which leads to fewer trap states associated with grain boundaries. This, although with a minimized migration of contaminate metals, can directly improve the photovoltaic performance. Also, a study of crystal structure confirms that a silicide layer is first formed, and this leads to the formation of distinct crystalline orientations of Si.

As the sputtering power decreases, the grain size increases due to a decrease in the sputtering rate. However, a decrease in the power causes a decrease in film thickness, which can be accounted for by an increase in sputtering time. As the sputtering power decreases, the photovoltaic performance improves. Therefore, by combining the parameters obtained from a study of metal thickness and sputtering power and time, a J_{sc} of 5 mA/cm^2 and V_{oc} of 0.24 V can be achieved. J_{sc} is now limited by photon absorption in the Schottky metal, lack of an antireflection coating, and non-passivation of grain boundaries. Ultimately, these films will be used for p-n junction or amorphous Si / μc-Si solar cells.

ACKNOWLEDGMENTS

Research was partially supported by the Air Force Office of Scientific Research, FA9550-07, monitored by Dr. Kitt Reinhardt.

REFERENCES

1. S. Hegedus, *Prog. Photovoltaics* **14** (5), 393-411 (2006).
2. J. Meier, R. Fluckiger, H. Keppner, A. Shah, *Appl.Phys. Lett.* **65** (7), 860-862 (1994).
3. J.K. Rath, *Sol. Energy Mater. Sol. Cells* **76** (4), 431-487 (2003).
4. C.H. Ji, W.A. Anderson, *IEEE Trans. Electron Devices* **60** (9), 1885-1889 (2003).
5. J.H. Choi, D.Y. Kim, S. S. Kim, S.J. Park, J. Jang, *Thin Solid Films* **440** (1-2), 1-4 (2003).
6. C. Hayzelden, J.L. Batstone, R.C. Cammarata, *Appl. Phys. Lett.* **60** (2), 225-227 (1992).
7. C.W.T. Bulle-Lieuwma, A.H. van Ommen, L.J. van IJzendoorn, *Appl. Phys. Lett.* **54** (3), 244-246 (1989).

Mater. Res. Soc. Symp. Proc. Vol. 1123 © 2009 Materials Research Society   1123-P05-14

Z.Q. Ma[1], B. He[1], J. Xu[2], L. Zhao[1], F. Li[1], N.S. Zhang[1], X.J. Meng[1], L. Shen[1] and C. Shen[1]

[1] SHU-SolarE PV Laboratory, Department of Physics, Shanghai University, Shanghai 200444, P.R. China

[2] State Key Laboratory of Advanced Technology for Materials Synthesis and Processing, Wuhan University of Technology, Hubei 430070, P. R. China

ABSTRACT

In order to fabricate AZO/SiO$_2$/p-Si heterojunction device and let it be an absorber of ultraviolet response cell. Zinc oxide (ZnO) thin films doped with aluminum (AZO) were deposited on p-Si(100) substrates covered with silicon dioxide (SiO2) by radio frequency magnetron sputtering. The optical and electrical properties of the Al doped - ZnO films were characterized by UV-VIS spectrophotometer, current-voltage measurement, and four point probe technique, respectively. The results show that the device is a typical tunneling diode for minority carrier and a strong obstructing effect from majority carriers. The potential rectifying behavior and photovoltaic characteristic is present at dark current and weak light illumination, respectively.

KEYWORDS: Photovoltaic, Optical properties, Electrical properties

INTRODUCTION

As shown in the previous work, semiconductor-insulator- semiconductor (SIS) diodes have certain features which make them more attractive for solar energy conversion than conventional Shottky, MIS, or other heterojunction devices [1]. For example, efficient SIS solar cells as indium tin oxide (ITO) on silicon have been reported, where the crystal structures and the lattice parameters of Si (diamond, a = 0.5431 nm), SnO$_2$ (tetragonal, a = 0.4737 nm, c = 0.3185 nm), and In$_2$O$_3$ (cubic, a = 1.0118 nm) show that they are not particularly compatible and thus are not likely to form good photovoltaic devices. However, the SIS structure is potentially more stable and theoretically more efficient than either a Schottky or a MIS structure. The origins of this potential superiority are the tunneling suppression of majority-carrier in the high potential barrier region of SIS structure, and the existence of the thin interface layer which minimizes the amount and the impact of the interface states. This results in an extensive selection of the p-n junction partner with a matching band gap in the front layer. In addition, the multi-function of the front semiconductor film could be served as an antireflection coating [2], a low-resistance window, as well as the collector of the p-n junction.

Furthermore, the semiconductor with a wide band gap as the top layer of SIS device can eliminate the surface dead layer which often occurs within the homojunction devices. On the other side, this absence of the light absorption of visible region in a surface layer can improve the ultraviolet response of the internal quantum efficiency (IQE). Among many transparent conductive oxides (TCO) of transition metals, ZnO:Al is one the best n-type semiconductor materials. It has low resistivity, high transmittance, optimized surface texture for light trapping, and wide bang gap of $E_g \approx 3.3eV$ [3]. Thus, in this report, we prepare AZO/SiO$_2$/p-Si device, as an attempt to study its opto-electronic conversion property and the I-V features as well.

EXPERIMENTAL DETAIL

For the purpose of fabricating SIS structure, p-type Si-(100) wafers were used as the substrates of the heterojunction to AZO. The wafers were prepared by a stand cleaning procedure, then they were dipped in 10% HF solution for one minute to remove native oxide layer. Finally, the wafers were dried in a flow of nitrogen.

By thermal evaporation, 1 μm-thick Al electrode was deposited on the back side. Then the samples were annealed at 500Ω for 20 min in air to form good ohmic contact and a very thin oxide layer (about 15~20Å) was grown on the p-Si surface.

The Al doped ZnO films were deposited on the oxidized silicon substrates in a RF magnetron sputtering system. The target was a sintered ceramic disk of ZnO doped with 2 wt % Al$_2$O$_3$ (purity 99.99%).The base pressure inside the chamber was pumped down to less than 5×10^{-4}Pa. Sputtering was carried out at a working gas (pure Ar) pressure of 1Pa. The Ar flow ratio was 30 sccm. The RF power and the temperature on the substrates were kept at 100W and 300°C, respectively. The sputtering was proceeded for 2.5 hours. The area is 2×2 cm^2.

The thickness of the AZO film was measured by a step profiler. The optical transmission of the films was measured by UV-VIS spectrophotometer. The electrical properties of Al doped ZnO films were characterized by four point probe. The current-voltage characteristics of the device was measured by Agilent 4155C semiconductor parameter analyzer (with probe station, the point of a probe is 5 μm).

RESULTS AND DISCUSSION

Optical and electric properties of AZO films

In order to learn the optical absorption and energy band structure of AZO films, the transmission spectrum of the Al doped ZnO film deposited on the glass substrate (sputtering time is 1 hour) was obtained. The thickness of AZO film is about 3560 Å. The average transmittance of the film is about 90% in the visible region. The absorption coefficient α was evaluated from the measurements of the optical transmittance T according to the following

34

relation [4]:

$$\alpha = \frac{1}{d}\ln\frac{1}{T} \qquad (1)$$

where d is the thickness of the AZO film.

The optical band gap of the AZO film is determined through the extrapolation of linear part of the absorption edge to $\alpha = 0$ in the relationship as:

$$(\alpha h v)^2 = A(E_g - h v) \qquad (2)$$

This result is basically obtained from the intrinsic absorption of electronic transition and the indirect measured value of about 3.24 eV is less than the crystalline zinc oxide, because the impurities in the films are responsible for the low energy tail, called an impurity gap [4].

The square resistance and the sheet resistivity of ZnO:Al film are measured by four-point probe to be 239.75 **H/□** and 0.0324 **H** cm, respectively.

I-V characteristics

A linear I-V behavior between the two electrodes on the surface of ZnO:Al film and Si indicates a good ohmic contact (not shown here). Fig.1 shows the current-voltage characteristic of the AZO/SiO$_2$-Si/Al heterojunction device measured at room temperature in the dark. Typical rectifying is observed. Basing on the dark current as a function of the applied bias, the corresponding diode resistance defined [5] as $R_D = (\frac{dI}{dV})^{-1}$ is derived and shown in Fig.2. It is obvious that the R_D – V relationship in several scopes is controlled by the various dark current transport mechanisms.

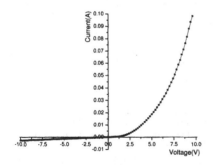

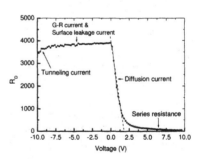

Fig.1 I-V curve of the Al/AZO/SiO$_2$/p-Si/Al heterojunction device in dark.

Fig.2. The variation of diode resistance via applied voltage (R$_D$-V).

35

The series resistance arose from ohmic depletion plays a dominant role when the forward bias is larger than 2.5 V. When the voltage is between 0 and 2 V, the resistance slightly increases with the diffusion current in the base region. When the inversion voltage increases from 0 to -5V, the leakage current and the generation-recombination (G-R) current in the surface layers restrain the increase trend of the dynamic resistance, which keeps the R_D – V curve in an unvaried state. The direct and indirect tunnel currents produced by higher inversion bias more than -5 V cause a slowly decrease of R_D.

The value of the ideality factor of the heterojunction is determined from the slop of the straight line part of the forward bias in log I - V characteristics, as shown in Fig3. At low forward bias (<2V), the typical value of the ideality factor and the reverse saturation current are 24.42 and 8.92×10^{-5}A, respectively. Using the standard diode equation as

$$I = I_0 \left(e^{\frac{qV}{nk_BT}} - 1 \right)$$

(3)

where q is the electronic charge, V is the applied voltage, k_B is the Boltzmann constant, n is the ideality factor and I_0 is the saturation current [6,7]. The values of n = 24.42 and $I_0 = 8.92 \times 10^{-5}$A are calculated through the least square multiplication fitting method. Fig.3 shows that the calculated result (using standard diode equation) is consistent with the measured curve.

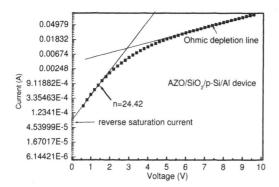

Fig.3 Illustrations of ideality factor n and reverse saturation current I_o derivation for the AZO/SiO$_2$/p-Si/Al device. All the values were obtained from the linear part of the I-V curve. The logarithmic scale is used for the current at forward bias.

Under the light illumination, the photovoltaic effect of AZO/SiO$_2$/p-Si based heterojunction device is shown in Fig.4 with the daylight lamp (weak light source). The photon-current response close to zero voltage is amplified and inserted in Fig.4. The open-circuit voltage is about 60 mV. The short-circuit current is about 13.8 μA. In contrast, the dark current reaches up to 3.94 mA when the reverse voltage is over10 V. Under reverse bias conditions, the photocurrent caused by the ZnO:Al surfaces exposing in daylight lamp is obviously lager than the dark current. The electron generated in the depletion region of the p-Si [8], is almost near the heterojunction edge. Light is absorbed in the p-Si side and the generated electrons and holes are drifted to AZO side and Si side, respectively. Thus the photocurrent is consequently obtained.

To understand the change of the dark- and photon-current, it should be noted that the role of intermediate layer is a significant factor. Majority carriers may be blocked from the tunneling by both the potential barrier within the interface and the band gap of AZO. For the minority carrier, the tunneling can occur via defects states at the two interfaces and the increased photoelectrical current may be the dominant tunnel transition via defects in AZO-Si region.

In a device such as the SIS diode, the two types of semiconductors are separated by a buffer layer with a high barrier. The insulator layer is sufficiently thin (10-20Å) so that current flows through the interface by tunneling mechanism. Since the tunneling process is sensitive to the precipitous degree of the barrier, it should be the charge transport behavior in the multi-layers semiconductor that determines the I-V characteristics.

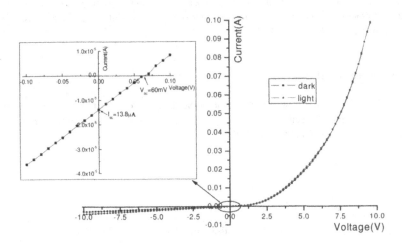

Fig.4 I-V characteristic of AZO/SiO$_2$/p-Si/Al device under dark and daylight lamp conditions.

Basically, the minority carriers (elections) which could enter into the AZO film are supplied by p-type silicon as an electron injector. The effective coupling current flows due to the interchange of the charge between the conduction and valence bands of the silicon by recombination-generation process. The degree of tunneling is a function of the thickness (d) of the interface layer. Where open-circuit voltage is given by:

$$V_{oc} = \frac{nkT}{q}[\ln(\frac{J_L}{A^*T^2}) + \frac{q\phi_B}{kT} + \sqrt{q\phi_T} \cdot d \ ]$$ (4)

Therefore, open-circuit voltage V_{oc} of SIS solar cell increases with the thickness of the interface layer. However, short-circuit current J_{sc} decreases with the thickness of the interface layer, which reduces the tunneling probability. It is believed that the appropriate thickness of the interface layer is about 20Å.

CONCLUSIONS

In summary, Al-doped n-type transparent conductive zinc oxide film has been successfully obtained by rf magnetron sputtering. The average transmittance of the AZO film is about 90% in the visible region. The optical gap E_g of ZnO:Al film is 3.24 eV. The square resistance and the sheet resistivity of the film are 239.75 H/□ and 0.0324 Hcm, respectively. The I-V curves of AZO/SiO$_2$/p-Si heterojunction device shows the good rectifying behaviors. The obvious photovoltaic feature is also illustrated under the daylight lamp illumination. The ideality factor n and the saturation current I_0 are 24.42 and 8.92×10^{-5}A, respectively. It is desired that the AZO/SiO$_2$/p-Si based heterojunction photovoltaic device will be a promising structure for the solar cell application in the future.

ACKNOWLEDGMENTS

This work was partly supported by Natural Science Foundation of China (No.60876045), Shanghai Leading Academic Discipline Project (S30105) and Innovation Foundation of Shanghai Education Committee (No.08YZ12).

REFERENCES

1. W. W. Wenas and S. Riyadi, Solar Energy Materials and Solar Cells 90, 3261 (2006)

2. D. Song, A.G. Aberle and J. Xia, Applied Surface Science 195, 291 (2002)

3. J. Lee, D. Lee, D. Lim and K. Yang, Thin Solid Films 515, 6094 (2007)

4. M. Selmi , F. Chaabouni, M. Abaab and B. Rezig, Superlattices and Microstructures 44, 268 (2008)

5. X. D. Chen, C. C. Ling, S. Fung and C. D. Beling, Applied Physics Letters 88, 132104

(2006)

6. S. Mridha, M. Dutta and D. Basak, Journal of Materials Science: Materials in Electronics 1573-482X (Online) (2008)

7. S. Mridha and D. Basak, Journal of Applied Physics 101, 083102 (2007)

8. J.Y. Lee, Y.S. Choi, J.H. Kim, M.O. Park and S. Im, Thin Solid Films 403-404,553 (2002)

Mater. Res. Soc. Symp. Proc. Vol. 1123 © 2009 Materials Research Society   1123-P06-07-F07-07

Dinesh Attygalle, Qi Hua Fan, Shibin Zhang, William B. Ingler, Xianbo Liao, Xunming Deng

Dept. Physics and Astronomy, Univ. Toledo, 2801 Bancroft Street, Toledo, OH 43606

ABSTRACT

To improve the cell efficiency of thin film solar cells textured back reflectors (BR) are widely used. This is particularly important in a-Si:H based solar cells due to low absorption coefficient at longer wavelengths. In this work we present a cost effective way to fabricate uniformly textured ZnO by using electrochemical methods. Further it was observed that Quantum Efficiency (QE) of shorter wavelengths also improved for highly textured ZnO BR. Together this resulted in more than 2mA increment in short circuit current density (J_{sc}) and 19% relative improvement in solar cell efficiency over sputter deposited BR. A possible mechanism responsible for the improved blue QE is also discussed.

INTRODUCTION

Light trapping is a method commonly used in thin film solar cells to enhance the long wavelength absorption. In a substrate type thin film solar cells using a textured back reflector does this [3]. Common back reflectors consist of a reflecting metal layer such as Ag or Al and a TCO layer such as ZnO. It was shown numerically by X. Yang [2] that the texturing on ZnO surface is more effective than the texture on metal.

ZnO can be deposited in numerous ways, such as RF sputtering, DC sputtering and PECVD. Alternatively the electrochemical deposition is also used for this purpose. A lot of research have been done to understand various aspects involved with ZnO electrodeposition [5,6]. The low cost associated with the electrodeposition techniques of ZnO attracted many researches and industrial community [1]. The non-uniformity was one of the major problem associated with electrochemical deposition of ZnO films, especially for large area applications like solar cells.

In this paper we discuss fabrication of ZnO films at high rate, using an improved electrochemical procedure, on stainless steel substrate, which is pre-coated with Ag/ZnO by sputter deposition. We demonstrate that, better uniformity and control over the ZnO texture can be achieved using this novel method. In addition, though it is not the prime purpose of BR, the favorable feature size helps to improve the blue response of the solar cell.

EXPERIMENT

The substrates were stainless steel sheets coated with Ag and ZnO layers by RF magnetron sputtering at the facility of Xunlight Corporation. The Ag and ZnO layers were about 300nm and 420nm thick, respectively. The BR sheet was then cut into 2"x4" pieces. The ZnO electrodeposition was done on top of the sputter deposited Ag/ZnO layers. The electrodeposition of ZnO can be directly done on Ag, how ever for comparison purposes Ag/ZnO sputter deposited substrates were used. Figure 1 shows the schematic diagram of the ZnO deposition setup.

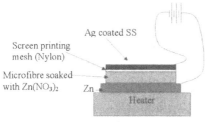

Figure 1 Improved ZnO electrodeposition setup

The anode is a 99.9+% pure Zn metal plate. The microfibre is soaked with 0.1M $Zn(NO_3)_2$ (aq) solution. The nylon mesh 361x361 from McMaster-Carr was used between microfiber and substrate to get better uniformity of deposition. The temperature was measured on the heater surface and stainless steel top surface by using a thermocouple. The temperature reading on the heater surface is set to 100 °C and the temperature on the stainless steel surface was measured to be 57±5 °C. Current was kept constant at 1mA/cm^2 for 20 minutes during the deposition. The samples were washed with DI water after the ZnO deposition.

AFM images were used to analyze the surface morphology. By using Shimadzu spectrophotometer attached with an integrating sphere, the total and diffused reflections of the BR's were measured. Thin film nc-Si single junction solar cells were deposited by VHF PECVD as in reference [4] by keeping the textured ZnO BR (electrodeposited ZnO on regular BR) and the control BR (regular BR without electrodeposited ZnO) side by side. The solar cell performance was measured by Oriel Solar simulator. The quantum efficiency of the solar cell is measured for wavelength range 380-1000 nm using a spectral response measuring system consisting of a monochromator and lock-in amplifier.

RESULTS / DISCUSSION

ZnO can be grown cathodically under a range of deposition conditions due to favorable Gibbs free energy of formation ($G_b^0 = -318\ kJmol^{-1}$) [7]. In electrochemical terms this gives rise to cathodic deposition potential range of −0.76 to 0.88 V versus the normal hydrogen electrode (NHE) at pH = 0 [8]. These conditions are determined by the standard potentials for the following reactions:

$$Zn^{2+} + \frac{1}{2}O_2 + 2e^- \rightarrow ZnO, \qquad E^0 = +0.88\ V\ versus\ NHE$$

$$Zn^{2+} + 2e^- \rightarrow Zn, \qquad E^0 = -0.76\ V\ versus\ NHE$$

Morphology of ZnO

In electrochemical deposition the growth depends on the ion diffusion in the metal-electrolyte interface and the bulk electrolyte. Use of microfiber and nylon mesh may have limited the free movement of ions in the bulk electrolyte. This results in a uniformly textured film as in Figure 2(c). From the analysis of the AFM data it can be clearly seen that rms roughness or the standard deviation of height of the surface change from 24.1nm to 81.5 nm, with the additional electrodeposited ZnO layer. As expected in both samples the roughness decreases after the solar cell deposition.

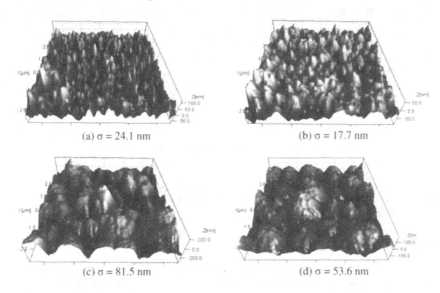

(a) σ = 24.1 nm   (b) σ = 17.7 nm

(c) σ = 81.5 nm   (d) σ = 53.6 nm

Figure 2 AFM images: (a) Control BR; (b) after the solar cell deposition on control BR; (c) textured ZnO BR (electrodeposited ZnO on control BR); (d) after the solar cell deposition on textured ZnO BR.

Optical Reflectance

A good light trapping BR effectively scatters the light incident onto its surface. The diffused reflectance measurements give a direct measurement of how much light is scattered. Figure 3 shows the increase on the diffused reflectance after the ZnO electrodeposition while the total reflectance remain same as regular BR, or in other words the total absorption by the BR is the same. The absorption of short wavelengths by the BR does not harm the solar cell performance as they are almost 100% absorbed by the intrinsic Si:H layers before reaching the BR.

43

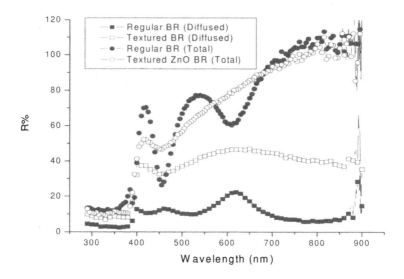

Figure 3 Diffused reflectance (normal incident) and Total Reflectance (8° incident angle) of BR samples with reference to Al mirror, measured with a spectrometer with integrating sphere attachment.

<u>Solar Cell Performance</u>

The main purpose of a back reflector is to scatter the long wavelength photons to trap them inside the solar cell. This increases the electron-hole pair generation, which ultimately results in higher current generation. From the J-V measurement it is clear that the textured ZnO BR has a higher current than the regular BR, see Figure 4(a). The effect of light scattering off the BR is clearly visible from the long wave region of the QE measurement, see Figure 4(b).

On the QE curve one noticeable thing is the increase in short wavelength region. To explain this phenomena we have correlate the device thickness and the surface roughness. The schematic of carrier transport for two cases are shown in the Figure 5. In the case of highly textured BR $\left(\sigma \geq \frac{absorber\ thickness}{2} \right)$, where σ is the rms roughness, the electrical conducting path from surface of the solar cell to the BR is shorter than the absorber layer thickness. For smooth or lightly textured BR this distance is almost equal to the absorber layer thickness. Since absorption coefficient of nc-Si for short wavelength is high, those photons generate carriers close to the front surface. The hole current through the device can be written as $J_p = -qD_p \frac{\partial p}{\partial x}\Big|_w$ [8], J_p is the hole current at $x=W$ surface, q is the effective charge, D_p is the hole diffusion coefficient, p is the carrier density and x is the distance. Since we use the identical conditions for solar cell deposition we can write the solution as

$$J_p = A \frac{1}{\sinh\left(W/L_p\right)}$$, where A is a constant to the two cases we consider and L_p is the hole

diffusion length.

In the highly textured BR effective device thickness W is reduced, which results in higher current through the device. This effect is particularly significant for short wavelengths. This mechanism explains why the blue QE response is higher for highly textured ZnO BR.

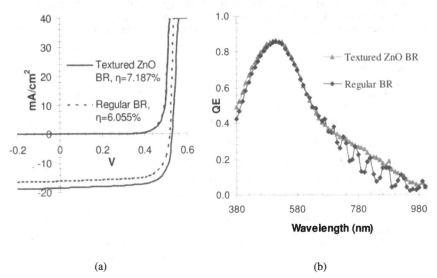

(a)   (b)

Figure 4 (a) J-V measurement of solar cells under AM 1.5 irradiation; (b) Quantum Efficiency of solar cells;

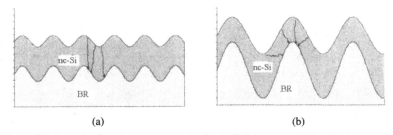

(a)   (b)

Figure 5 Schematic of carrier transport paths for (a) lightly textured BR; (b) highly textured BR.

45

CONCLUSIONS

A cost effective technique to deposit uniformly textured ZnO was developed. The solar cell back reflectors made using this technique were successfully tested. The highly textured nature of these back reflectors work as an effective scattering surface and make effective light trapping device for nc-Si solar cell. Further the highly textured nature of these devices changes the effective length of electric conducting path inside the device, which enables higher spectral response in short wavelength region.

ACKNOWLEDGMENTS

The authors would like to thank Xunlight Corperation for supplying the experiment materials. This work was supported by U.S. Department of Energy Grant DE-FG-OSGO85025 and National Renewable Energy Laboratory Thin Film Partnership Program grant ZXL-5-44205-06.

REFERENCES

1. K. Saito, M. Sano, S. Okabe, S. Sugiyamab and K Ogawa; Solar Energy Materials & Solar Cells 86 (2005) 565–575.
2. X. Yang, *"Study of Transparent Conductive Oxides and Back Reflectors for Amorphous and nano-Crystalline Silicon Based Thin Film Solar Cells"*, Ph.D Dissertation, Unversity of Toledo (2007).
3. X. Deng and E. Schiff, *"Amorphous Silicon Related Solar Cells"*, a chapter in *Handbook of Photovoltaic Engineering*, ed. A. Luque & S. Hegedus, John Wiley & Sons, Ltd. (2002).
4. Deng, X.; Cao, X.; Ishikawa, Y.; Du, W.; Yang, X.; Das, C.; Vijh, A. IEEE WCPEC (2006).
5. M. Izaki and T. Omi, J. Electrochem. Soc, 143, No 3, (1996) 53-55.
6. M Izaki and Takashi Omi; Appl. Phys. Lett., 68, (1996) 2439-2440.
7. Daniel Lincot; Thin solid films, 487, (2005) 40-48.
8. T. Pauporte and D. Lincot, Appl. Phys. Lett., 75 (1999) 2861
9. S. M. Sze, *"Physics of Semiconductor Devices"*, 2nd Ed., John Wiley & Sons, Ltd. (1981) 54.

Mater. Res. Soc. Symp. Proc. Vol. 1123 © 2009 Materials Research Society   1123-P07-11

Studies on Backside Al-Contact Formation in Si Solar Cells: Fundamental Mechanisms

Bhushan Sopori,[1] Vishal Mehta,[1] Przemyslaw Rupnowski,[1] Helio Moutinho,[1] Aziz Shaikh,[2] Chandra Khadilkar,[2] Murray Bennett,[3] and Dave Carlson[3]

[1] National Renewable Energy Laboratory, Golden, CO 80401, USA

[2] Ferro Electronic Materials, Vista, CA 92083, USA

[3] BP Solar, Frederick, MD 21703, USA

ABSTRACT

We have studied mechanisms of back-contact formation in screen-printed Si solar cells by a fire-through process. An optimum firing temperature profile leads to the formation of a P-Si/P$^+$-Si/ Si-Al eutectic/agglomerated Al at the back contact of a Si solar cell. Variations in the interface properties were found to arise from Al-Si melt instabilities. Experiments were performed to study melt formation. We show that this process is strongly controlled by diffusion of Si into Al. During the ramp-up, a melt is initiated at the Si-Al interface, which subsequently expands into Al and Si. During the ramp-down, the melt freezes, which causes the doped region to grow epitaxially on Si, followed by solidification of the Si-Al eutectic. Any agglomerated (or sintered) Al particles are dispersed with Si. Implications on the performance of the cell are described.

INTRODUCTION

The contact formation of most commercial Si solar cells is done by co-firing the cells with screen-printed front and back contacts. Typically, a gridded front-contact pattern of Ag-based ink is directly applied on the antireflection coating of SiN:H, whereas the backside has a blanket Al-based contact. The firing process, also called fire-through contact metallization, performs the following functions: (i) On the front contact, the glass frit dissolves the antireflection coating of SiN:H and allows Ag to react with Si. This reaction involves participation of a solvent metal that leaches from the glass frit and helps to lower the eutectic point of Si-Ag alloy; (ii) Diffuses hydrogen from the SiN:H-Si interface into the bulk of the cell and passivates impurities and defects; and (iii) Produces an interaction of Si and Al to form a deep back-surface field (BSF) and sinters unreacted Al to form a low-resistance contact. It is important to understand how each of these functions of this complex process can be optimized.

This paper describes our investigations on the formation of the back contact, and how most of the requirements of a good back contact can be met by a suitable process. We will first describe requirements of a back contact and then discuss our experiments and results.

Requirements of a Good Back Contact

These requirements may be divided into electronic and optical categories. The electronic properties include: (i) a BSF for minority-carrier reflection, (ii) a uniform low-resistance ohmic contact, and (iii) a smooth, dimple-free surface of Al. Optically, the back contact must be reflecting with very little absorption so that it contributes effectively to light trapping. In addition to these, the process of making the back contact should be compatible with efficient gettering of impurities. In commercial solar cell fabrication, the primary consideration is given to the electronic properties, while optical reflection and gettering are assumed as byproducts, because there is some concern whether all the requirements can be met simultaneously. Indeed, our detailed analyses indicate that the performance of the back contact of most commercial cells is below optimum.

The back-contact formation relies heavily on alloying of Si and Al to produce a controlled stratified structure. To obtain the desired electronic properties of the back contact, the kinetics of Si-Al melting and re-solidification of various phases during the firing cycle must be controlled. The next section briefly discusses requirements and issues with each region.

Back-surface field

The purpose of the BSF is to provide a minority-carrier reflector that can lower the loss of photo-generated carriers at the back surface [1–6]. As is well known, the BSF arises from the band bending at the P^+–P region. Figure 1(a) illustrates the geometry of the back contact and

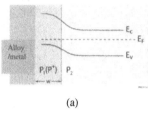

(a)

Figure 1. (a) Illustration of band bending within the P^+-P region, (b) calculated effective surface recombination velocity at a BSF.

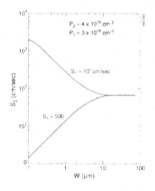

(b)

shows a band diagram of the backside of a typical boron-doped solar cell. One approach to determine the effectiveness of the BSF is to determine the surface recombination velocity (SRV) of the P^+–P interface, which depends on many factors including the thickness of the P+ region. Figure 2(b) shows the calculated SRV at the back contact as a function of the thickness of the P+ region [7]. These results are given for two values of S1. Here, S1 is the recombination velocity at the alloy-P+ interface and S2 is the effective recombination velocity due to the entire contact. It is clear that the width of the P+ region should be about 10 μm for a good BSF. Thus, it is important to determine how the cell should be processed to create a uniform, 10-μm-wide P+ region in conjunction with a good front contact.

Low-resistance contact

Low-resistance contact formation necessitates that a well-formed, ohmic contact between Si and Al be produced beyond the P^+ region. An ideal contact would consist of Al/P^+/Si (as shown in Fig. 1(a). However, as shown in the actual cross-section of a cell later in the paper, a eutectic composition accompanies the formation of the P^+ region. Because the eutectic composition has a higher sheet resistivity compared to Al, this can lead to increased series resistance of the cell. An obvious way to mitigate this issue is to increase the thickness of the Si-Al eutectic. We also show that the back contact typically leaves a region of disconnected sintered Al particles in a glass matrix. These regions can also contribute to the series resistance of the finished cell.

Other properties

In addition to being a minority-carrier reflector, a back contact must also be a good optical reflector to enhance light trapping within the cell [8]. The co-firing process should also effectively getter the impurities in the cell. The details of these will be discussed elsewhere.

EXPERIMENT

Experiments were done on single wafers and multicrystalline (mc) Si solar cells. These cells were fabricated on 124 mm x 124 mm, textured wafers using commercial screen-printed front and back contacts. The wafers were single crystal, single-side polished with Al deposited by electron-beam deposition either on the polished side or rough side. The firing was done in an optical furnace in which the cell is illuminated with a tailored optical flux profile, and the temperature of the wafer/cell is monitored to determine local temperatures at various critical areas such as under the metallization and away from it. All single-crystal samples were illuminated from the Si side.

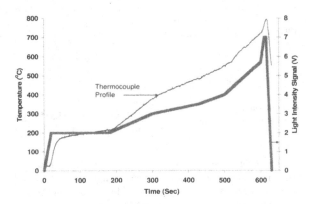

Figure 2. A typical firing profile along with measured cell temperature for Si solar cell.

A typical firing profile for a solar cell is shown in Fig. 2. The ramp-up segment accomplishes the following: bake the cell to drive out organics, produce a molten glass to etch the SiN:H antireflection coating, melt Si-Al on the back side, and sinter Si-Ag on the front side. The ramp-up is followed by a short constant temperature. This maximum temperature determines the maximum concentration of Al in the Si melt, which controls the maximum doping of the P+ region (see the discussion). In the ramp-down part, the molten regions of both the front and back contact solidify. Because the single-crystal samples had evaporated Al, the ramp-up portion of the profile was eliminated.

Characterization Techniques

The processed samples were characterized by a variety of techniques. Scanning electron microscopy (SEM) was used for high-resolution imaging of the processed contacts and for dopant profiling to study the BSF formation. Scanning Kelvin probe microscopy (SKPM) was used to generate potential profiles to evaluate the formation of the BSF and to determine local conductivity. Secondary-ion mass spectrometry (SIMS) profiles were used to measure Si diffusion in Al and profiles of Al after firing. Cross-sectioning was performed by a new technique [9]. This technique produces a highly flat cross-section over a large length of the cell that allows one to perform statistically meaningful analysis. Solar cells were characterized by dark and illuminated current-voltage (I-V) analyses.

RESULTS AND DISCUSSION

Optical microscopy of our cross-sectioned samples provided valuable information on melting of Al and formation of Si-Al alloy. Figure 3 shows an optical micrograph of the back side of an optimally fired cross-sectioned Si solar cell with screen-printed contacts. This figure shows that the back contact consists of: (i) Al-doped region that has grown epitaxially on the Si surface. The thickness of this region is about 15 μm. The interface waviness is due to texturing of the cell. (ii) Alloyed region (Si-Al eutectic) shows typical lamellar growth consisting of Si-rich and Al-rich phases. (iii) A granular region consisting of sintered Al particles in a matrix of glass.

Interface roughness due to texturing of the wafer. Note: The texture of the interface is retained.

(a)   (b)

Figure 3. A cross-section of an optimally fired solar cell showing back-contact region.

Figure 4 (a) and (b). SEM images of different areas of same optimally fired multi-Si cell.

It may be noticed that some of the Al particles have melted and combined with neighboring particles. The size of the Al particle is between 5 and 20 μm. Because our technique allows large samples to be cross-sectioned, we have observed that large variations exist in the thickness of alloyed regions and in the size of the sintered Al particles.

Figure 4 shows SEM images of the back contact of a fired mc-Si solar, taken in dopant contrast. It shows a P^+ region, as well as the alloyed eutectic and sintered Al regions. The thickness of this region is only 3 μm and varies along the length of the interface. Figure 4b is an image of another section of the same cell showing a significant variation in the thickness of the alloyed layer and of P^+ region and that some correlation exists between the thicknesses of these regions.

Figure 5a shows an SEM image of a back contact and its potential profile (Fig 5b) under different bias conditions obtained by SKPM. A voltage of 110 mV appears across this P^+-P region. Figure 6 is an optical micrograph showing a very important feature of the alloy thickness variation; the thickness of the alloy is largest at the valley of the texture and a minimum at the peak of the texture. This indicates that during the formation of Si-Al alloy, the melt initially fills up the texture valley. Any excess melt spills over the adjacent valleys and then builds up. The depth of the P^+ region is generally larger at thicker alloyed regions.

Figure 5. (a) An SEM image (b) corresponding SKPM image (potential profile).

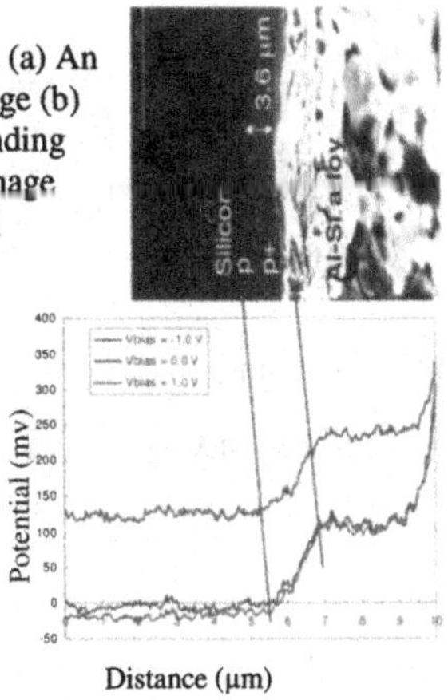

Figure 6. Optical micrograph of a c/s cell, showing thickness variations of the alloyed region corresponding to the textured profile of the Si.

Our results show that the melting and solidification of the back Al contact agrees well with the binary phase diagram of Al and Si [10]. It is rather remarkable that the general (qualitative) features of the phase diagram, which is applicable to a thermal equilibrium process, are indicated in our results (which may not be an equilibrium process). Indeed, there are variations imposed by the departures from a phase diagram. These departures arise from the following:

1. We have observed that the firing process involves a very rapid diffusion of Si into Al. This diffusion controls the initial composition and location of the initial melt near the Si-Al interface. Figure 7 shows a SIMS profile of Si diffused into Al caused by a short firing profile; the thickness of Al was kept at only 1 micron to show that the diffused Si can accumulate at the Al-air interface. In a screen-printed cell, the diffused Si can disperse between melted Al particles. Figure 8 shows an SEM (Fig. 8a) and EDX images of Si (Fig 8b) and Al (Fig 8c). In Fig. 8b, one can see that Si has diffused deep into Al beyond the alloyed region. The presence of Si around Al particles will cause an increase in the series resistance of the cell.

2. Melting of Al on a Si surface produces a very nonuniform melt because of high surface tension (large contact angle) and balling up into Al spheres. This phenomenon will lead to pockets of melt of different sizes, concomitantly producing P^+ and Si-Al eutectic regions of different thicknesses (as seen in Figs. 4 and 6). We have found that Si diffusion into Al can be used to increase the adhesion between the Al melt and Si, thereby producing a uniform melt, and hence, a uniform contact.

3. During ramp-down, the melt cools and tries to regrow epitaxially over Si. This doped layer of Si follows the topology of the Si surface. On further cooling, an outside layer of Al freezes, which traps a liquid layer between solid Al (particles) and Si (with regions of soft glass). As the melt reaches the eutectic point, the entire melt is expected to freeze. It may be recognized that beyond the eutectic point Si will continue to diffuse into Al-Si and will pile up at interstitial spaces. Another important significance of diffusion of Si is that it competes with Al for oxygen that may be available at the interface, and it prevents a strong oxidation of melted Al particles to allow them to join with each other.

51

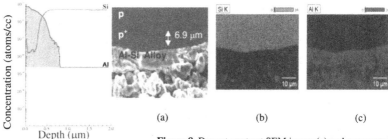

Figure 7. SIMS profile of Si diffusion in Al.

Figure 8. Dopant-contrast SEM image (a) and corresponding EDX images of (b) Si and (c) Al.

CONCLUSION

An optimum firing temperature profile leads to the formation of a P-Si/P⁺-Si/ Si-Al eutectic/unmelted Al at the back contact of a Si solar cell. Variations in the interface properties were found to arise from Al-Si melt instabilities. Experiments show that melt formation is strongly controlled by diffusion of Si into Al. This affects the series resistance of the cell.

Based on the above results, a desirable profile should encourage initial Si diffusion. This will initiate a melt very close to the eutectic point of 577°C. Next, the temperature should be raised to about 775°–800°C to ensure an adequate amount of Al concentration in the melt to form heavily doped P⁺ upon solidification. The duration of the peak temperature strongly controls the total thickness of P+ and the eutectic regions. However, a slow cooling from peak temperature to the eutectic point is desirable to ensure a large P+ region.

REFERENCES

1. J. D. Alamo, J. Eguren, and A. Luque, *Solid-State Electron.* **24**, 415 (1981).
2. C. Khadilkar, S. Kim, A. Shaikh, S. Sridharan, and T. Pham, *Tech. Digest. of International PVSEC- 15*, (2005).
3. A. Kaminski, B. Vandelle, A. Fave, J.P. Boyeaux, L. Q. Nam, R. Monna, D. Sarti, and A. Laugier, *Solar Energy Materials & Solar Cells* **72**,373 (2002).
4. S. Narasimha, A. Rohatgi, and A. W. Weeber, *IEEE Trans. on Electron Devices* **46**(7), 1363 (1999).
5. F. Huster and G. Schubert, 20th European Photovoltaic Solar Energy Conference and Exhibition, Barcelona, 2DV2.48, 6–10 June (2005).
6. C.H. Lin, S. Y. Tsai, S. P. Hsu, and M. H. Hsieh, *Solar Energy Materials & Solar Cells* **92**, 986 (2008).
7. B. Sopori, V. R. Mehta, P. Rupnowski, D. Domine, M. Romero, H. Moutinho, B. To, R. Reedy, M. Al-Jassim, A Shaikh, N. Merchant, and C. Khadilkar, Proc. 22nd European Photovoltaic Solar Energy Conference, Milan, 841(2007).
8. M. Cudzinovic and B. Sopori, Proc. 25th IEEE PVSC, 501 (1996).
9. B. Sopori, V. Mehta, N. Fast, H. Moutinho, D. Domine, B. To, and M. Al-Jassim, Proc.17th Workshop on Crystalline Silicon Solar Cells & Modules: Materials and Processes, Vail, Colorado, 222 (2007).
10. J. L. Murray and A. J. McAlister, *Bull. Alloy Phase Diagrams* **5**, 74 (1984).

Mater. Res. Soc. Symp. Proc. Vol. 1123 © 2009 Materials Research Society   1123-P02-01

Lirong Zeng, Peter Bermel, Yasha Yi, Bernard A. Alamariu, Kurt A. Broderick, Jifeng Liu,
Ching-yin Hong, Xiaoman Duan, John Joannopoulos, and Lionel C. Kimerling
Massachusetts Institute of Technology, Cambridge, MA, 02139

ABSTRACT

The major factor limiting the efficiencies of thin film Si solar cells is their weak
absorption of red and near-infrared photons due to short optical path length and indirect bandgap.
Powerful light trapping is essential to confine light inside the cell for sufficient absorption. Here
we report the first experimental application of a new light trapping scheme, the textured photonic
crystal (TPC) backside reflector, to monocrystalline thin film Si solar cells. TPC combines a one-
dimensional photonic crystal, i.e., a distributed Bragg reflector (DBR), with a reflection grating.
The near unity reflectivity of DBR in a wide omnidirectional bandgap and the large angle
diffraction by the grating ensures a strong enhancement in the absorption of red and near-infrared
photons, leading to significant improvements in cell efficiencies. Measured short circuit current
density J_{sc} was increased by 19% for 5 μm thick cells, and 11% for 20 μm thick cells, compared
to theoretical predictions of 28% and 14%, respectively.

INTRODUCTION

Thin film silicon solar cells are already emerging as a leading next generation
photovoltaic technology, due to their potential of low cost. However, currently they suffer from
low efficiencies due to weak absorption of long wavelength photons. Effective light trapping
schemes are crucial to enhance absorption of long wavelength photons and increase cell
efficiency.

Traditional light trapping schemes are based on geometrical optics elongating optical
path length by scattering at the textured front surface [1] and reflecting at the back surface with
an aluminum reflector, which has a maximum reflectivity of ~80% when light is incident from Si.
Even the combination of an ideally roughened front surface and a lossless rear reflector is
theoretically forbidden from enhancing path lengths by more than 50 times the cell thickness [2],
while the best experimental result is closer to a factor of ten [3].

Recently we demonstrated the efficiency enhancement in thick Si solar cells using a new
light trapping structure, the textured photonic crystal backside reflector [4], which combines a
one-dimensional photonic crystal as a distributed Bragg reflector (DBR) and a reflection grating.
The design was optimized with coupled wave theory [5] and scattering matrix method [6], and
broader theoretical studies on the light trapping properties of photonic crystals were undertaken
[7]. We now report the first experimental integration of this structure with optimized parameters
onto monocrystalline thin film Si solar cells. This light trapping technique, developed from the
foundations of wave optics theory, allows us to effectively target the longer wavelengths
requiring trapping, using the combination of an extremely high omnidirectional reflectivity
mirror effective over a wide stopband of DBR and strong large angle diffraction from the grating.
Optical path length can be effectively changed from the thickness of the cell to its lateral
dimension, leading to significantly enhanced absorption.

SIMULATION

Fig. 1 is the schematic of a solar cell integrated with the textured photonic crystal back reflector, which is composed of a rectangular grating and a DBR stack. There are three types of control cells in our study: reference cells without back structure, where the grating and DBR in Fig. 1 are replaced by a 500 nm thick SiO₂ for electrical isolation from the Si substrate; "DBR-only" cells where the grating in Fig. 1 is eliminated;   and "grating-only" cells where the DBR in Fig. 1 is replaced by 500 nm thick SiO₂ for electrical isolation.

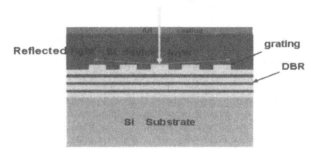

Figure 1. Schematic of a thin film Si solar cell integrated with textured photonic crystal backside reflector consisting of a reflection grating and a distributed Bragg reflector

To quantify the light trapping capability of textured photonic crystal, scattering matrix method [8] was used to simulate the absorption spectrum and calculate the cell efficiency. The DBR and grating parameters are the optimal ones found in Ref. 6. The antireflection coating (ARC) is composed of an 8 nm thick underlying thermal oxide and a Si_3N_4 layer around 70 nm thick. For the simulation comparison, 100% carrier collection efficiency was assumed. Figure 2 (a) and (b) displays the calculated absorption spectra of Si solar cells with 5 and 20 μm thick device layers. The as-calculated spectra showed sharp peaks due to thin film interference effect and grating diffraction. In order to compare with experimental results later, the spectra were smoothed out with a moving average method which preserves the area under the curves; furthermore, 20.4% shadowing from electrodes is taken into account. Clearly, at short wavelengths, the absorption spectra for cells with different back structures overlap, as short wavelength photons will be completely absorbed by the thin Si film and will not be able to reach the back reflector. At longer wavelengths, however, cells with back structures display higher and wider absorption curves.

The collective effect of the enhancement in absorption of long wavelength photons due to DBR and grating can be computed by the improvement in short circuit current density J_{sc}, calculated from

$$J_{sc} = q \int_{\lambda=300nm}^{1200nm} A(\lambda)s(\lambda)d\lambda \qquad (1),$$

where q is the electronic charge, $A(\lambda)$ is the absorption, and $s(\lambda)$ is incident solar photon flux density from the AM1.5 spectrum. The calculated J_{sc} for 5 μm thick cells with the

54

aforementioned four back structures are: reference cell-15.8 mA/cm^2, grating-only cell-16.9 mA/cm^2 (9.1% relative enhancement); DBR-only cell-17.4 mA/cm^2 (10.0% relative enhancement); and DBR+grating cell-20.2 mA/cm^2 (28.3% relative enhancement). For 20 μm thick DBR+grating cells, the relative J_{sc} enhancement over the reference cell is 13.7%. The relative power conversion efficiency enhancement is almost the same as the relative J_{sc} enhancement, due to the ideal diode and 100% carrier collection assumptions.

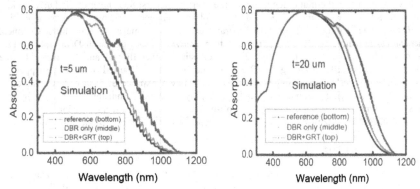

Figure 2. Simulated absorption spectra of Si solar cells with differing back structures and active layer thicknesses. (a) 5 μm; (b) 20 μm. 20.4% shadowing is considered.

EXPERIMENT

SOI solar cell design and fabrication

To prove the absorption enhancement capability of TPC, top-contacted monocrystalline thin film Si solar cells integrated with the textured photonic crystal back reflector were successfully fabricated using Si-on-insulator (SOI) material through an active layer transfer technique. Lateral junctions were adopted. Monocrystalline Si was used to eliminate the complication of materials quality issues associated with deposited thin films and to make the optical effect obvious. If the active layer is changed to polycrystalline Si, the relative efficiency enhancement should not change because the optical properties will remain virtually the same. The active layer thicknesses were 5 and 20 μm, and four different back structures were involved for each cell thickness: none (reference structure), DBR-only, grating plus wavy DBR (TPC), and grating plus flat DBR (TPC). The DBR stack was composed of 6 bilayers of SiO$_2$ and Si. The grating parameters and DBR film thicknesses used were the optimal values used in simulation. If the DBR stack is deposited after grating etching, the stack film will adopt the waviness of the underlying grating, as shown in the transmission electron microscopy image in Ref. 9, which was thought beneficial to light trapping by causing more diffraction compared to the flat stack film. To verify this, control samples with grating plus flat DBR were made by planarizing the first layer of DBR stack film with chemical mechanical polishing after the grating etch. The cell size was 4.3 mm^2, with 20.4% shadowing due to metallization for all samples. To simplify processing, no electrode optimization was performed.

In order to form textured photonic crystal on the backside of a monocrystalline Si thin film, an active layer transfer technique was used. The starting material was SOI wafers. The first step was to do a blanket ion implantation to form a back surface field. Gratings with periods of around 300 nm were then patterned with interference lithography and etched with reactive ion etching. Then a thin thermal oxide was grown as part of the first SiO$_2$ layer of the DBR stack. 6 pairs of SiO$_2$/Si were then deposited using plasma enhanced chemical vapor deposition (PECVD). The grating, DBR and ARC parameters are the same as in simulation. Chemical mechanical polishing (CMP) was performed to planarize the top surface of DBR, followed by fusion bonding to a new handle wafer. The bonded pair was then flipped, and the original handle wafer was removed by grinding, CMP and plasma etching; and buried oxide was stripped. A double-layer antireflection coating was formed on the newly exposed Si surface, consisting of an 8 nm thick underlying thermal oxide and a Si$_3$N$_4$ layer around 70 nm thick, deposited by low pressure chemical vapor deposition. Lateral p-i-n junctions were then formed by ion implantation. After contact etching, 2 μm thick Al-2% Si was deposited by sputtering for metallization.

Solar cell characterization

Dark I-V characteristic and external quantum efficiency (EQE) measurements were carried out with an H20 IR Jobin Yvon monochromator coupled to a Hewlett-Packard 4145A semiconductor analyzer. J_{sc} was measured using a sun simulator.

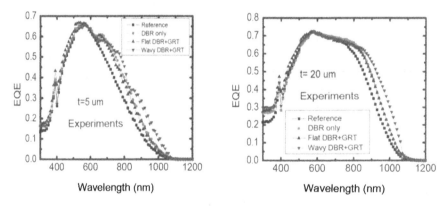

Figure 3. Measured external quantum efficiency for solar cells with differing back structures and active layer thicknesses. (a) 5 μm; (b) 20 μm. The discontinuity at λ=400 nm is due to the introduction of a filter to remove the 2nd harmonic from the monochromator light source.

RESULTS AND DISCUSSION

Dark I-V measurements showed good rectifying behavior for all the solar cells, with leakage current at a few nA/cm^2. Figure 3 (a) and (b) depicts the measured EQE for cells with different thicknesses and back structures. Figure 3 (a) clearly shows that at $\lambda<640$ nm, the EQE spectra of all cells almost completely overlap for 5 μm Si film thickness, but for $\lambda>640$ nm the EQE of cells with different back structures diverges. The reference cell has the lowest EQE and smoothest curve, the flat DBR plus grating cell and DBR-only cell have one bump on EQE curves, and the wavy DBR plus grating cell has the highest EQE and at least three bumps, corresponding to strongly enhanced absorption. Figure 3 (b) for 20 μm thick cells shows that while all the curves overlap for $\lambda<740$ nm, cells with back reflector structures display higher and wider EQE spectra for longer λ compared to the reference cell, and cells with wavy DBR plus grating have the highest EQE. An inspection of Figure 2 and 3 reveals that the measured EQE closely matches the simulation results in trend and magnitude.

The joint effect of the enhancement in EQE of long wavelength photons due to the back reflectors is embodied by the increase of short circuit current density J_{sc}. Fig. 4 depicts the simulated (considering shadowing effect) and experimental J_{sc} for solar cells with different back structures and cell thicknesses. The measured J_{sc} is quite close to simulation, with their ratio varying from 81.6% for 5 μm thick "flat DBR+grating" cell to 97.4% for 20 μm thick DBR-only cell. This discrepancy comes from electronic recombination not accounted for in the simulation. As expected, cells with wavy DBR+grating show the highest J_{sc}, followed by flat DBR+grating, DBR-only, and the reference cell. Experimentally, the 5 μm thick wavy DBR+grating cell achieves 18.9% J_{sc} enhancement compared to the reference cell, while the theoretical prediction is 28.3%; and the corresponding enhancement for the 20 μm thick cell is 11.3%, with the simulated value of 13.7%.

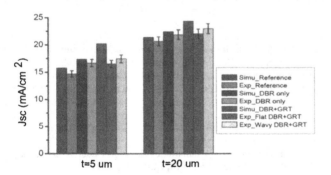

Figure 4. Simulated and experimental J_{sc}'s with differing back structures for 5 and 20 μm thick solar cells. For each cell thickness, from left to right, the bars correspond to simulated or experimental J_{sc} for cells with different back structures, as listed in the legend from top to bottom. Error bars are shown on experimental results.

The higher J_{sc} of wavy DBR+grating compared to flat DBR+grating verifies our previous assumption that the extra periodicity in the DBR film plane does enhance light trapping. Therefore, it might be beneficial to make the grating less square to increase the waviness of DBR. It is also worth mentioning that using two-dimensional gratings in textured photonic crystal instead of 1D gratings can increase cell efficiency further because light can be diffracted in two perpendicular directions, as confirmed in our recent publication [7].

CONCLUSION

In conclusion, we have experimentally demonstrated that textured photonic crystal backside reflector combining reflection grating and distributed Bragg reflector can significantly enhance absorption in the red and infrared spectral regime in thin film solar cells, as predicted by simulation, due to the combined effects of high DBR reflectivity and strong grating diffraction. We report 19% and 11% respective increases in J_{sc} in 5 and 20 µm thick monocrystalline Si solar cells made from SOI wafers through an active layer transfer technique. The textured photonic crystal back reflector we developed can be applied directly to monocrystalline and polycrystalline Si solar cells, and its principle is broadly applicable to other materials systems. DBR can be conveniently deposited using PECVD. With large scale production methods for grating such as nanoimprint lithography [10] and better surface passivation, textured photonic crystal is expected to significantly boost thin film solar cell efficiency.

ACKNOWLEDGEMENTS

We thank J. Michel for equipment calibration and helpful discussions on data analysis, X. Sheng for help on solar cell measurements, and the Nanoscience @ UNM facility, a member of the National Nanotechnology Infrastructure Network for assistance on interference lithography.

REFERENCES

1. M. A. Green, *Silicon Solar Cells, Advanced Principles and Practices* (Center for Photovoltaic Devices and Systems, University of New South Wales, Sydney, 1995), p.92.
2. E.Yablonovitch and G. Cody, IEEE Trans. Electron Devices **ED-29**, 300 (1982).
3. J. Nelson, *The Physics of Solar Cells* (Imperial College, London, 2003), pp.279 and 282.
4. L. Zeng, Y. Yi, C. Hong, J. Liu, N. Feng, X. Duan, and L. C. Kimerling, Appl Phys. Lett. **89**, 111111 (2006).
5. N.-N. Feng, J. Michel, L. Zeng, J. Liu, C. Hong, L. C. Kimerling, and X. Duan, IEEE Trans. Electron Devices **54**, 1926 (2007).
6. L. Zeng, P. Bermel, Y. Yi, N. Feng, B. A. Alamariu, C. Hong, X. Duan, J. Joannopoulos, and L. C. Kimerling, Mater. Res. Soc. Symp. Proc. **974**, CC02-06 (2007).
7. P. Bermel, C. Luo, L. Zeng, L. C. Kimerling, and J. D. Joannopoulos, Opt. Express **15**, 16986 (2007).
8. D. Whittaker and I. Culshaw, Phys. Rev. B. **60**, 2610 (1999).
9. L. Zeng, Y. Yi, C. Hong, B. A. Alamariu, J. Liu, X. Duan, and L. C. Kimerling, Mater. Res. Soc. Symp. Proc. **891**, 251(2006).
10. L. J. Guo, Proc. SPIE **5734**, 53 (2005).

Mater. Res. Soc. Symp. Proc. Vol. 1123 © 2009 Materials Research Society   1123-P07-09

Ultrafast Laser Textured Silicon Solar Cells

Barada K. Nayak, Vikram Iyengar and Mool C. Gupta[1]

Charles L. Brown Department of Electrical and Computer Engineering, University of
Virginia, Charlottesville, VA 22904

ABSTRACT

A novel ultrafast laser texturing method has been developed to produce arrays of
nano/micro surface textures in silicon. Laser processing conditions have been optimized
for achieving appropriate optical and electronic properties for photovoltaic applications.
The textured silicon surfaces absorb greater than 99% of incident light over the entire
solar spectrum and the material appears complete black to bare eye. Textured silicon
surfaces are characterized for surface morphology and optical properties. Chemical
etching and thermal annealing steps have been performed to remove laser slag and
induced defects. Finally, we report the total and external quantum efficiency results on
photovoltaic devices fabricated on the textured silicon wafers, which can be further
improved.

INTRODUCTION

Efficient light energy absorption to cause carrier generation in a photoactive material is
one of the most important requirements for enhancing photovoltaic energy conversion in
solar cells. Therefore, high efficiency solar cell devices require good reflection control to
minimize optical loss at the front surface. This is typically achieved with a combination
of surface texturing (using wet chemical process) and anti-reflection coating (by
deposition of Si_3N_4 layer using PECVD method). Usually industry uses two types of
texturing procedure:
(a) Anisotropic etching for mono-crystalline silicon using alkaline solutions like NaOH
and KOH at 70-80 ° C with addition of isopropanol [1]. This etching process produces
randomly distributed pyramids and suffers from following shortcomings: lack of pyramid
size control, poor reproducibility, presence of untextured regions, need for adequate
surface preparation, and requirement of accurate control of temperature and isopropanol
concentration. While this method is applicable for mono-crystalline materials, it could
not be applied to multi-crystalline materials due to the anisotropic nature of the etchant.
(b) Isotropic texturing for multi-crystalline silicon using acidic mixture of HF and HNO_3
and organic additives [2]. This method relies on the residual saw damage that causes
uneven etching of the surface that result in the texture formation. However, acidic
texturing has several issues for multi-crystalline wafers due to: requirement of saw
damage for effective texturing; accurate surface preparation, etc. Other methods of
texturing like reactive ion etching, mechanical texturing and photolithography and wet

---

[1] Corresponding author E-mail: mgupta@virginia.edu

etchings has been proposed but are not cost effective. On the other hand, lasers are unique energy sources that could texture surface by selectively removing material by ablation process. Since laser ablation is an isotropic process, texturing could be achieved irrespective of the crystallographic orientation of material surface. Several groups have demonstrated that laser could be used to texture the surface by creating grooves [3,4] or pits [5]. Previously we have reported that an ultrafast laser could be used to create arrays of self-assembled nano/micro surface textures on various materials [6-9]. These nano/micro textures trap light very efficiently (over 99 % from UV to IR) irrespective of the angle of incidence and state of polarization. Also lasers are increasingly used for photovoltaic manufacturing. Here for the first time we report photovoltaic device results fabricated on the ultrafast laser textured silicon surfaces, where further improvements in efficiency are possible.

EXPERIMENT

Boron doped FZ silicon (100) of resistivity 2 Ω cm is diced into small square chips of size 2 cm x 2 cm. One of the chips is put on a stage inside a vacuum chamber (with base pressure ~ 1 mbar) mounted on a computer controlled X-Y stage. The chamber is rinsed with SF_6 at least twice and then backfilled with 400 mbar SF_6. The samples are exposed to 0.6 mJ pulses of 800 nm wavelength and 130 fs pulse duration at a repetition rate of 1 kHz from a regeneratively amplified Spectra Physics Ti-sapphire laser system. The laser beam is focused along the normal onto the sample surface by a 200 mm focal length lens and the laser fluence is adjusted by using a Glan laser calcite polarizer. In order to expose an area bigger than the laser spot size, the samples are raster scanned under the laser beam using the motorized X-Y stage. Scanning is adjusted such that laser lines overlap suitably to generate a uniform surface texture over the silicon surface.

Laser textured wafers are first cleaned using a chemical etching procedure (NaOH solution rinse followed by Isotropic etching (HNO_3:CH_3COOH: HF: 30: 10:4)) to remove laser slag and smoothen the textured cones. The textured wafers are further cleaned by IMEC process [10] for removal of organic and metal contamination. Wafers are then annealed at 900 C for 30 minutes in an inert (90% Ar + 10% H_2) environment to passivate any laser induced defects. In order to form an n+p junction, commercially available phosphorus doped oxide (P509 from Filmtronics) solution is spun on the texture surface. After spin coating, the wafers are baked at 200 °C for 15 minutes before they are inserted into a diffusion furnace. Doping is carried out in a N_2:O_2:3:1 ambient at 900 °C for 15 minutes. The spin on dopant film is retained after diffusion and front electrode area is patterned using photolithography. Aluminum (700 nm) is sputtered on to the back of the device (p-side) and Ti/Ag (50 nm/700 nm) is e-beam evaporated onto the front side of the device (n+ side) for the ohmic contact. The front electrode shadowing is around 6% of the device surface area. Lastly, the samples undergo a forming gas anneal at 450 °C for 30 minutes. Devices are then edge isolated before further characterization.

Surface morphology of the textured surfaces are analyzed using a scanning electron microscope (SEM) (Zeiss SUPRA 40), reflectance measurements are carried out using a spectrophotometer, I-V measurements are carried out using a class B solar simulator from

PV Measurements and photovoltaic efficiency is calibrated using a commercial solar cell calibrated by NREL for 1.5 SUN (100 mW/cm^2) illumination.

RESULTS AND DISCUSSION

Fig. 1 shows the SEM image of as-laser textured (a) single crystalline and (b) multicrystalline (c) textured and chemically etched silicon surface. It is evident from the image that the size of the conical microstructures are ~ 10 μm tall with base diameter of ~ 5 μm tapering down to ~ 100 nm and separated by ~ 6 μm from each other. Figure 1 (a) and (b) indicate the laser slag that is produced by redeposition of ablated material. A thin disordered layer over the conical structures is also produced during rapid laser melting and resolidification process [11]. Figure 1 (c) demonstrate that most of the laser slag and amorphous layer are successfully removed by employing a chemical cleaning followed by thermal annealing mentioned in the experimental section. The blackness of the textured material is retained even after the chemical etch (see the inset picture of a solar cell of size 1 cm x 1 cm fabricated on a textured surface in figure 2).

| (a) | (b) | (c) |

Figure 1: SEM images of (a) as-laser textured CZ silicon (b) as-laser textured multi-crystalline silicon (c) after NaOH and isotropic etching. Scale bar for all the images is 20 μm.

Figure 2 demonstrates the impact of surface texturing on the reflectivity of silicon. It is clear that the laser textured surface absorb the entire visible spectrum of the electromagnetic radiation compared to its un-textured counter part. We have also observed that reflectivity turns out to be less than 10 % up to 15 μm wavelength in the IR region of the spectrum [9]. The total scattering from the textured surface is less than 1% measured using an integrating sphere.

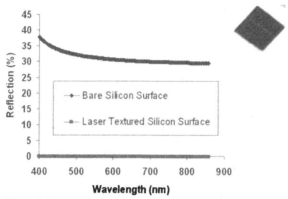

Wavelength (nm)

Figure 2: Light reflection from polished bare silicon and laser textured silicon surfaces after chemical etching and thermal annealing process. Inset shows the photograph of a diced device fabricated on textured silicon wafer.

Figure 3 (a) and (b) show the dark and 1 SUN illuminated I-V data for the fabricated device. It is evident from the short circuit current value that the textured surface has efficiently trapped incident light. However, the low open-circuit voltage and fill factor have caused the reduction in the over all efficiency. The low open-circuit voltage can be attributed to high reverse saturation current as seen in figure 3 (a). The high recombination current could result from high doping levels for the emitter region and/or residual laser induced defects. We believe that optimization of doping and appropriate etching for this textured material could reduce the recombination current.

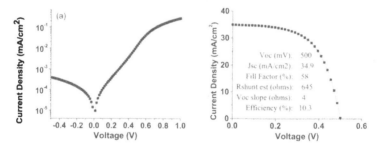

Figure 3: Current-voltage characteristics of textured silicon solar cell at 1 SUN illumination. Left side curve indicates dark I-V characteristics.

Another possibility for poor fill factor could be attributed to high series resistance. Scanning electron microscope images (figure 4) clearly indicate the porosity in the front metal film deposited for electrical contact. The devices showed external quantum efficiency (EQE) at an average of 75 % over the entire visible spectrum. In spite of the

62

high absorption in this material, the reduction in overall EQE could be due to the porosity and/or non-uniformity of the metal layer leading to poor charge collection.

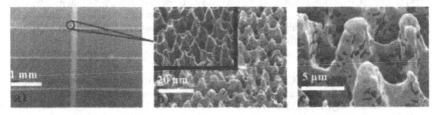

Figure 4: SEM images of (a) device showing front finger electrodes (b) a higher magnification region of (a) indicating an area at the crossing of bus bar and finger. The dark line separates the metallized region (white area) from the unmetallized region (c) Higher magnification image of metallized region of (b) showing the porosity/discontinuity in the metal film deposited for ohmic contact.

CONCLUSIONS

In conclusion, we have demonstrated for the first time that solar cell devices of over 10% efficiency could be fabricated using ultrafast laser textured silicon substrates. It should also be noted that the fabricated devices are not optimized for best surface passivation, doping profile and do not have a back surface field. So further improvements in device efficiency could be realized with optimization of fabrication process. The laser texturing process is unique due to the following reasons: (a) textured surfaces absorb around 99 % of incident radiation over the entire solar spectrum (b) process is highly reproducible and is relatively fast (c) the process produces same micro textures on multi-crystalline silicon wafers.

Post chemical and thermal treatments on as-textured surfaces seem to have removed most of the laser induced defects that is evident from the high photocurrent. Thick metallization is required to overcome the observed porosity in the deposited metal film. With appropriate finger grid geometry, surface passivation, back surface field and better charge collection, ultra fast laser texturing can yield high efficiency photovoltaic cells.

ACKNOWLEDGEMENTS
We thank NSF and NASA-Langley for their financial support for this project.

REFERENCES
[1] D. L. King and M. E. Buck, "Experimental Optimisation of an Anisotropic Etching Process for Random Texturisation of Silicon Solar Cells", Proc. 22nd IEEE Photovoltaic Specialists Conf., pp. 303-308 (1991).
[2] D. Sarti, Q. N. Le, S. Bastide, G. Goaer, and D. Ferry, "Thin industrial multicrystalline solar cells and improved optical absorption", Proceedings of the 13th European PV Solar Energy Conference, pp. 25–28, 1995.
[3] J. C. Zolper, S. Narayanan, S. R. Wenham and M. A. Green, "16.7 % efficient, laser

textured, buried contact polycrystalline silicon solar cell", Appl. Phys. Letts., 55 (22), 2363 (1989).

[4] L. Pirozzi, G. Arabito, F. Artuso, V. Barbarossa, U. Besi-Vetrella, S. Loreti, P. Mangiapane, E. Salza, Selective emitters in buried contact silicon solar cells: Some low-cost solutions", Solar Energy Materials & Solar Cells 65, 287-295 (2001).

[5] Malcolm Abbott, and Jeffrey Cotter, "Optical and Electrical Properties of Laser Texturing for High-efficiency Solar Cells", Prog. Photovolt: Res. Appl. ; 14:225–235, (2006).

[6] Barada K. Nayak, Mool C. Gupta and Kurt W. Kolasinski, "Spontaneous formation of nano-spiked microstructures in germanium by femtosecond laser irradiation", Nanotechnology, 18, 195302. (2007).

[7] Barada K. Nayak, Mool C. Gupta and Kurt W. Kolasinski, "Ultrafast-laser-assisted chemical restructuring of silicon and germanium surfaces", Applied Surface Science, 253, 6580, (2007).

[8] Kurt W. Kolasinski, Margaret E. Dudley, Barada K. Nayak, and Mool C. Gupta, "Pillars formed by laser ablation and modified by wet etching", Invited Paper – Proceedings of SPIE, The International Society for Optical Engineering, Vol. 6586, Prague, Czech Republic, (2007).

[9] Mool C. Gupta and Barada K. Nayak, "Systems and methods of laser texturing and crystallization of material surfaces", International Patent Application Serial No. PCT /US2006/049065), filed on December 21, 2006 in the U.S. Patent and Trademark Office.

[10] M. M. Heyns, T. Bearda, I. Cornelissen, S. De Gendt, R. Degraeve, G. Groeseneken, C. Kenens, D. M. Knotter, L. M. Loewenstein, P. Mertens, S. Mertens, M. Meuris, T. Nigam, M. Schaekers, I. Teerlinck,W. Vandervorst, R. Vos, and K.Wolke, "Cost-effective cleaning and high-quality thin gate oxides," IBM J. Res. Develop., vol. 43, no. 3, p. 339, May 1999.

[11] C. H. Crouch, J. E. Carey, M. Shen, E. Mazur, F. Y. Genin, " Infrared absorption by sulfer-doped silicon formed by femtosecond laser radiation, " Appl. Phys. A, 79, 1635 (2004).

Thin Film Polycrystalline Materials
and Devices

Mater. Res. Soc. Symp. Proc. Vol. 1123 © 2009 Materials Research Society 1123-P05-18

Effects of Growth Parameters on Surface-Morphological, Structural and Electrical Properties of Mo Films by RF Magnetron Sputtering

Shou-Yi Kuo[1,2], Liann-Be Chang[1,2], Ming-Jer Jeng[1,2], Wei-Ting Lin[1,2]
Yong-Tian Lu[3] and Sung-Cheng Hu[3]
[1] Department of Electronic Engineering, Chang Gung University,
[2] Green Technology Research Center, Chang Gung University,
259 Wen-Hwa 1st Road, Kweishan, Taoyuan 333, Taiwan
[3] Chemical Systems Research Division, Chung-Sung institute of Science & Technology

ABSTRACT

This work reports on the fabrication and characterization of Mo thin films on soda-lime glass substrate grown by reactive RF magnetron sputtering. Film thickness was measured by α-step surface profiler. The structural properties and surface morphology were analyzed by x-ray diffraction (XRD), atomic force microscope (AFM) and scanning electron microscopy (SEM). Electrical properties were measured by four-point probe. It was found that the growth parameters, such as argon flow rate, RF power, film thickness, have significant influences on properties of Mo films. The strain on films revealed the complicated relationship with the working pressure, which might be associated with micro structures and impurities. In order to improve the adhesion and electricity, we adopted a two-pressure deposition scheme. The optimal thickness and sheet resistance are 1 μm and 0.12 $\Omega/\square$. The mechanisms therein will be discussed in detail. Furthermore, we also investigated the diffusion property of Na ion of double Mo films sputtered on soda-lime glass. Our experimental results could lead to better understanding for improving further CIGS-based photovoltaic devices.

INTRODUCTION

Molybdenum (Mo) is commonly used as back contacts for $Cu(In,Ga)Se_2$ (CIGS) thin film solar cells because it has good electrical properties and is an inert, mechanically durable substrate during the absorber film growth [1]. A wide variety of other materials such as W, Ta, Nb, Cr, V, Ti, and Mn have also been investigated without any improvement compared to Mo, and often lower cell efficiency due to chemical reactivity. CIGS thin film growth occurs at the temperatures above 500 ^{0}C and the thermally induced extrinsic stresses in the glass-Mo structure may cause bending or mechanical distortion of substrate. During the formation of CIGS films, Na ions might diffuse from the soda-lime glass substrate through the Mo back contact into the absorber layer. The diffusion of Na into the absorber film depends on the deposition conditions of the Mo back contact. Thus, to fabricate high-efficiency CIGS solar cells, we need to understand the properties of sputtered Mo films. The deposition parameters and process play a key role in obtaining a layer with the appropriate properties. Extensive research has been done on the deposition of Mo thin films by dc sputtering. In this article, we report the surface morphology, structural, electrical and strain properties of Mo thin films prepared by RF magnetron sputtering.

EXPERIMENT

Molybdenum (Mo) thin films were deposited on soda-lime glass (SLG, 10x10 cm^2) substrates with a standard thickness of 1 μm by RF magnetron sputtering system using high purity Mo target (99.99%, 6 inches diameter). The target-to-substrate distance was 6 cm, and the substrate was rotated at 15 rpm for thickness uniformity. Before depositing the Mo thin films, the glass substrates were burnished by Al$_2$O$_3$ solution and ultrasonically cleaned in a detergent bath, followed by acetone, isopropanol and DI water. Afterward the substrate surfaces were checked by UV-254 spectrophotometer. The sputter chamber was evacuated to a base pressure of 4x10^{-6} Torr by a turbomolecular pump. The working pressure was varied from 1.0 to 10 mTorr, while the sputtering power was fixed at 300 W. The film thickness was kept constant at 1 μ m, and all films were deposited at room temperature. In second case, we have designed a two-power and two-pressure deposition in which we first sputtered a thin layer of "high argon pressure" Mo to serve as an adhesion layer, followed by the deposition of "low argon pressure" Mo to achieve low sheet resistance and free strain. Film thickness was measured by α-step surface profiler. The structural property and surface morphology were analyzed by x-ray diffraction (XRD), atomic force microscope (AFM) and scanning electron microscopy (SEM). Electrical properties were measured by four-point probe. Na diffusion was measured by Energy Dispersive Spectrometer (EDS).

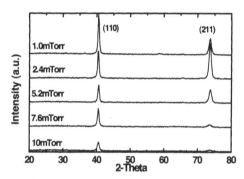

Figure 1. Typical XRD patterns of Mo thin films sputtered under various argon pressure.

RESULTS AND DISCUSSION

Study as a function of the argon pressure

The X-ray diffraction spectra of Mo films grown at different working pressures are shown in the Figure 1. From figure 1, there are two main peaks with orientation along the (110) and (211) direction. As the argon flow rate was increased, the peak intensity of Mo(110) diffraction plane decreased, which were probably caused by increased oxygen content or growth energy decreased. Figure 2 shows the variation of the full width at half maximum (FWHM) of the (110) diffraction peak, as a function of the working pressure. As we increase the working pressure, the FWHM values monotonically increase as well. The tendency indicated that Mo films sputtered at low pressures have higher crystallinity than films deposited at high pressures. This can be correlated with the dense microstructures obtained for films grown at low pressures. The average particle size of the Mo films can be calculated from the broadening of the corresponding (110) x-ray peaks using Scherrer formula [2]:

$$L = \frac{K \cdot \lambda}{B \cdot \cos(\theta_B)} \tag{1}$$

where L is the crystallite size, K is the Scherrer constant (0.9 in this case), λ is the wavelength of x-ray (CuKα=0.154 nm), and B is the FWHM of diffraction peak at θ_B. The grain size was found to decrease from 16.5 to 11.5 nm as the argon pressure increases, with a proportional trend obviously for the FWHM as shown in figure 2. The FWHM increases with deposition argon pressure, and it leads to a decrease on Mo grain sizes, which were probably caused by increased oxygen content or growth energy decreased.

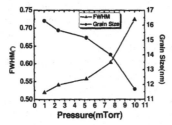

Figure 2. FWHM and grain size of Mo thin films as a function of argon pressure.

In order to further confirm the surface morphological properties of RF-sputtered Mo films. Figure 3(a) and 3(b) shows the surface roughness of the Mo films for various values of the working pressure. The surface roughness of the films increased with the working pressure. In addition, AFM studies showed good agreement with XRD measurements : As the average RMS roughness decreased from 7.27 to 1.9 nm with increasing working pressure, and the grain size decreased as well.

Figure 3. AFM images of Mo films sputtered at (a) 2.4 mTorr and (b) 7.6 mTorr.

Figure 4 shows the intrinsic stress value of the Mo films as a function of the working pressure. The residual stress calculations were made by strain formula from XRD data. Using

Bragg's law, the inter-planar lattice spacing, $d_{(110)}$, was calculated. The strain on films is then calculated using the following formula:

$$Strain(\%) = (\frac{\Delta a}{a}) \times 100\% \qquad (2)$$

where 'a' is the lattice constant (for the Mo reference, a = 0.31472 nm), and the main parameter to determine whether the strain is tensile or compressive [3]. From figure 4, it is obvious that the variation of residual stress can be divided into four regions. Starting from low pressure, the strain increases first with increasing pressure and reaches a maximum value at 2.4 mTorr (region I); it then decreases with further increase in pressure (region II) until it goes into a compressive strain at the turning point around 7.5 mTorr (region III and IV). The origin of the strain profiles in sputtered Mo films may be related to several factors, including voids, oxygen or argon impurities, and crystallographic flaws [4]. In region I, the decrease in tensile strain with decreasing pressure is attributed to the less frequent formation of voids resulting from the impact of higher energetic particles at low pressures. At low enough pressure, one can see that the films would probably also reach a compressive strain (the limit in our case is due to the design of the deposition chamber) [5]. In region II, the increase of pressure increases the frequency of gas phase collisions, reducing the kinetic energy of sputtered neutral atoms and reflected neutrals bombarding the growing films. The decrease in tensile strain with decreasing incident kinetic energy (i.e., increasing working pressure) may be due to a micro-structural change from a densely packed network with atomic scale voids to a micro-columnar structure, associated with the incorporation of impurities such as oxygen and hydrogen [6–8]. In region III, the decrease in tensile strain is high enough and induces a compressive strain. In region IV, the decrease in compressive strain, which is probably caused by decreased growth energy or the structure were destroyed with the incorporation of impurities.

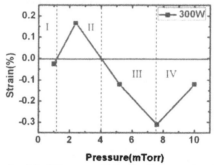

Figure 4. Strain of the RF sputtered Mo thin films as a function of argon pressure.

Figure 5 shows the variation of electrical resistance of the sputtered Mo films with respect to the working pressure. The electrical resistance of the Mo films increased continuously with increasing working gas pressure except for the lowest working pressure. Thin films grown at lowest argon pressure were unable to pass the tape test for adhesion. The lowest sheet resistance was ~0.18 $\Omega/\square$ and the electrical properties were remarkably improved with decreasing argon pressure [9,10]. These structural and electrical properties are summarized in table 1.

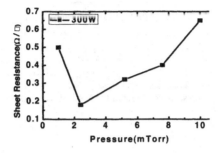

Figure 5. Sheet resistance as a function of deposition pressure.

Table 1. XRD, AFM , electrical properties and adhesion measurements as a function of argon pressure for sputtered Mo thin films.

Pressure (mTorr)	Strain(110) (%)	FWHM(110) (deg)	Orientation (110)	RMS	Adhesion Test	Resistance ($\Omega/\square$)
1.0	-0.03	0.52	0.70	X	Fail	0.5
2.4	0.166	0.54	0.44	3.16	Pass	0.18
5.2	-0.12	0.56	0.56	7.27	Pass	0.32
7.6	-0.31	0.60	0.82	2.62	Pass	0.4
10	-0.12	0.72	0.78	1.9	Pass	0.65

Mo bi-layer process

The adhesion of the film to the substrate was studied by simple adhesive scotch tape test. As mentioned in our experimental results, the Mo films sputtered at a single pressure can't simultaneously possess low sheet resistance and good adhesion. In order to circumvent this problem, we have designed a two-pressure deposition scheme in which we first sputter a thin layer of "high argon pressure" Mo to serve as an adhesion layer, followed by the deposition of "low argon pressure" Mo to achieve low sheet resistance. The method was used to make the CIGS layer. We used the strain free parameter (1.5 mTorr) for Mo top layer, because the thermal expansion coefficient of CIGS which was bigger than Mo material. The procedure for fabricating the Mo layer was similar to that described earlier, except that the Ar pressure was varied during the deposition, as shown in figure 6. This recipe has been used to routinely fabricate nominally 1.0 μm thick with sheet resistances in the range 0.12-0.18 $\Omega/\square$ that pass the tape test for adhesion. Besides, the Na diffusion content from soda-lime glass to the CIGS layer is about 0.52~0.76 at.%.

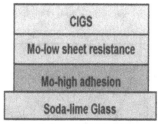

Fig 6. Cross-sectional view of the Mo bi-layer structure.

CONCLUSIONS

In summary, RF magnetron sputtering was used to deposit molybdenum thin films on soda-lime glass. Our experimental results indicated that the working pressures strongly influence the physical, electrical and structural properties of the films. The preferred orientation of films was along the (110) plane and the diffraction intensity decreases with increasing gas pressure. Except for the lowest working pressure, the sheet resistance of the Mo films increased continuously with increasing working gas pressure. In order to improve the adhesion and electricity, we adopted a two-pressure deposition scheme. The optimal thickness and sheet resistance are 1 μm and 0.12 Ω/□. The value for Na diffusion content is about 0.52~0.76 at% and the control of the Na-content in the layer is underway. These results are important for potential market in CIGS solar cells.

ACKNOWLEDGMENTS

The authors would like to gratefully acknowledge partial financial support from the Green Technology Research Center of Chang Gung University.

REFERENCES

1. M. A. Contrearas, B. Egaas, K. Ramanthan, J. Hiltner, A. Swartzlander, F. Hasoon and R. Nou_, Prog. Photovolt. Res. Appl. 7, 311 (1999).
2. Kasai N and Kakudo M 2005 *X-ray Diffraction by Macromolecule* (New York: Springer) pp 364–5
3. JCPDS card 42-1120 Mo cubic a = 0.31472 nm.
4. S. G. Malhotra, Z. U. Rek, S. M. Yalisove and J. C. Bilello, *J. Vac. Sci. Technol.* A 15 345 (1997).
5. H Khatri and S Marsillac. *J. Phys. : Condens. Matter* 20 (2008) 055206.
6. T. Yamaguchi and R. Miyagawa R, *Jpn. J. Appl. Phys.* 30 2069 (1991).
7. I. Blech and U. Cohen, *J. Appl. Phys.* 53, 4202 (1982).
8. H. Windischmann, R. W. Collins and J. M. Cavese, *J. Non-Cryst. Solids* 85, 261 (1986).
9. J. L. Alleman, H. Althani, R. Noufi, H. Moutinho, M. M. Al-Jassim, F. Hasoon: "Dependence of the Characteristics of Mo Films on Sputter Conditions", NCPV Program Review. Meeting, pp. 239-240 (2000).
10. J. H. Scofield, A. Duda, D Albin, B.L. Ballard, P.K. Predecki. *Thin Solid Films* 260, 26-31 (1995).

Mater. Res. Soc. Symp. Proc. Vol. 1123 © 2009 Materials Research Society 1123-P07-01

Improvement of the interface quality between Zn(O,S,OH)$_x$ buffer and Cu(InGa)(SeS)$_2$ surface layers to enhance the fill factor over 0.700

K. Kushiya, Y, Tanaka, H. Hakuma, Y. Goushi, S. Kijima, T. Aramoto, Y. Fujiwara, A. Tanaka, Y. Chiba, H. Sugimoto, Y. Kawaguchi and K. Kakegawa

CIS Development Group, New Business Development Div., Showa Shell Sekiyu K.K., 123-1, Shimo-kawairi, Atsugi, Kanagawa, JAPAN

ABSTRACT

For realizing the proof of mass production capability or a move toward a GW/a production, 16%-efficiency project has been started setting the target of each parameter as V_{oc}: 0.685 V/cell, FF: 0.735 and J_{sc}: 31.8 mA/cm^2. Up to FY2008, the target of each parameter independently has been achieved except the efficiency. All of our research works is currently focused on the FF in order to achieve the FF of over 0.73 consistently by adjusting the two resistances (R_{sh} and R_s) in the monolithically integrated 30cmx30cm-sized circuits. To improve the FF, double buffer structure with CBD-Zn(O,S,OH)$_x$ and MOCVD-ZnO is proposed and the thickness is adjusted by optimizing the R_s and the R_{sh}. As the result, FF of over 0.7 has been achieved for the first time in our CIS R&D since FY1993.

INTRODUCTION

Showa Shell Solar K.K., a 100 % subsidiary of Showa Shell Sekiyu K.K., has started up the second Cu(InGa)(SeS)$_2$ (CIS)-based thin-film PV plant of annual production capacity of 60 MW/year this November. The plant is currently in the initial test run stage. While, the first plant of annual production capacity of 20 MW/year, which started up in October, FY2006, is improving both the performance and the production yield. By the middle of 2009, two manufacturing plants will be in full operation. This means that we could successfully transfer our CIS-based thin-film Photovoltaics (PV) technology developed under PV-R&D projects by the New Energy and Industrial Technology Development Organization (NEDO) from FY1993 to FY2005 at the Atsugi Research Laboratory (ARL) with an at most 2 MW/year R&D line, which could achieve the average efficiency of 12 % on a 30cmx120cm-sized glass substrate. In the first plant, the substrate size was employed twice larger, 60cmx120cm size to reduce the production cost. The production know-how, lessons and operational experiences obtained in the first plant were all transferred to the three-times larger second plant combining with our baseline manufacturing technology. The CIS-based thin-film PV module from the production line of the first plant demonstrated the module (aperture-area) efficiency of 13.1 % measured by the National Institute of Advanced Industrial Science and Technology (AIST) as shown in Table 1. This achievement was recognized that our baseline manufacturing technology developed in the R&D stage at the ARL was successfully transferred to the first plant. In Table 1, the device structure is common and was prepared by the baseline manufacturing process described in detail elsewhere [1,2], i.e. the stacked structure of ZnO:B (BZO) window by a metal-organic chemical vapor deposition (MOCVD) technique / Zn(O,S,OH)$_x$ buffer by a chemical-bath deposition

(CBD) technique / Cu(InGa)(SSe)$_2$ (CIGSS) surface layer on a Cu(InGa)Se$_2$ (CIGS) absorber by sulfurization after selenization (SAS) process / Mo base electrode by sputtering / soda-lime glass as a substrate.

Table 1 Our best efficiencies.

Achieved by	Efficiency [%]	Aperture Area [cm^2]	Output [W$_p$]	Measured by
Showa Shell Sekiyu	15.2	855	13.0	ARL
Showa Shell Sekiyu	13.6	3456	47.0	ARL
Showa Shell Sekiyu	**13.4**	**3459.3**	**46.3**	**NREL**
Showa Shell Solar	**13.1**	**7128**	**93.1**	**AIST**

Since the performance of 12% average efficiency is not enough to compete with the commercialized polycrystalline-Si PV modules, improvement in the efficiency up to 16 % on a 30cmx30cm-sized substrate was set as the target in the R&D, which would be transferred to the third plant and could be translated into the performance of 14% average efficiency on a 60cmx120cm-sized substrate. To achieve this target, R&D works have been focusing on the interfaces in the stacked structure of constituent thin films, especially improvement in the pn hetero-interface between CIGSS surface and Zn(O,S,OH)$_x$ buffer layers. These efforts led to the fill factor (FF) of over 0.700 for the first time in our CIS R&D. In this study, some approach and consideration to improve the FF would be reported.

EXPERIMENT

The purpose of this study was to improve the FF investigating each interface in the device as shown in Fig. 1. For this purpose, we have focused on the interface between Zn(O,S,OH)$_x$ buffer and CIGSS surface layer, although actual device structure is more complicated and has more layers, such as MOCVD-BZO window/ MOCVD-ZnO layer/ CBD-Zn(O,S,OH)$_x$ buffer/ CIGSS surface layer/ CIGS absorber/ MoSe$_2$ (or Mo(SSe)$_2$) layer/ Mo/ glass substrate.

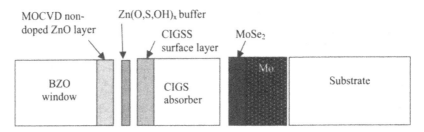

Fig. 1 Actual device structure in our CIS-based thin-film solar cells.

74

The reasons why we selected the process and the materials in our baseline process are summarized in Table 2. In this study, our baseline process was adjusted in order to solve the remaining question for a long time, "Why our CIS-based thin-film solar cells can not attain the FF of over 0.700?".

Table 2 Reasons why we selected the process and the materials in our baseline process.

Constituent thin film	Reasons of selection
Soda-lime glass as a substrate	Temperature durability and price.
Silica as an alkaline barrier	Availability and price.
Mo as a metal base electrode	Corrosion resistance on selenium, availability and price.
Cu-Ga alloy/In as a metal precursor	Wider band-gap with Ga, cost reduction due to the recycle of end-of-life sputtering targets.
Diluted H_2Se gas for Selenization Diluted H_2S gas for Sulfurization	Reduction of Se consumption was effective, because our SAS process was estimated roughly 1/40 of coevaporation case with solid Se source. Surface passivation with S to reduce the defects and double graded band-gap structure with Ga and S.
CBD-Zn(O,S,OH)$_x$ buffer	We did not want to apply CdS and ZnO-based materials are appropriate to increase the J_{sc} in the short wave-length region.
MOCVD-ZnO:B window	Sputtered-AZO (or GZO) TCO window could not make a good junction with CBD-Zn(O,S,OH)$_x$ buffer because of a physical process. MOCVD as a chemical process works well to make a better junction with our CBD-Zn(O,S,OH)$_x$ buffer. Our CIS-based absorber with a graded band-gap structure needs more incident light in a longer wave-length region up to 1200 nm. We could develop the in-line type MOCVD machine successfully and transfer it to the commercialization stage.

DISCUSSION

To achieve the 16 % efficiency on a 30cmx30cm-sized substrate, improvement in the FF was understood to be critical if 16 % efficiency has been achieved by the following formula as Eff. (16 %) = V_{oc} (0.685 V/cell) x J_{sc} (31.8 mA/cm^2) x FF (0.735). However, "Why we can not attain the FF of over 0.700?" has been a major question in our CIS R&D for a long time. To accelerate the technology development to achieve the 16 % efficiency, CIS R&D switched the substrate size from 30cmx120cm to 30cmx30cm in FY2006. We could achieve a 15 % milestone in FY2007 mainly as the result of improvement in the FF over 0.700 for the first time in our CIS R&D works as shown in Fig. 2.

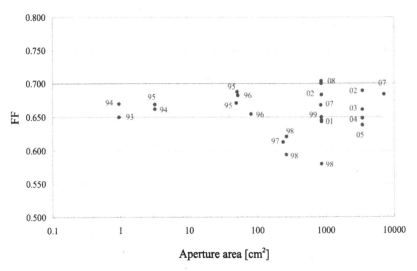

Fig.2 FF records of Showa Shell from FY1993 to FY2007. (Number in the graph corresponds to the year achieved the FF.)

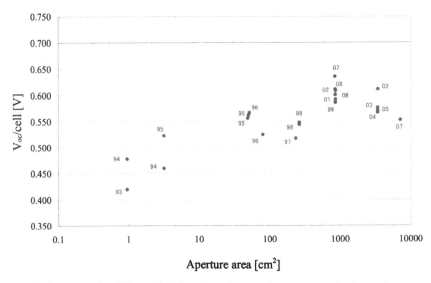

Fig.3 V_{oc} records of Showa Shell from FY1993 to FY2007. (Number in the graph corresponds to the year achieved the V_{oc}/cell.)

We did not focus on the interface of Mo base electrode and CIS-based absorber, which was well controlled by forming the $MoSe_2$ (or $Mo(SSe)_2$) and due to its crystallinity [3]. In our manufacturing process, the series resistance (R_s) as a slope at the V_{oc} and the shunt resistance (R_{sh}) as a slope at the J_{sc} were employed, because such simple engineering yardstick was effective to evaluate the interface quality. Therefore, approach to reduce the R_s and increase the R_{sh} was performed mainly focusing on the pn hetero-junction between CIGSS surface layer and $Zn(O,S,OH)_x$ buffer. To increase the R_{sh}, the sulfurization step of the SAS process was carefully adjusted employing the "delta $T_{sul-sel}$ concept" as a process control parameter [4]. V_{oc} was intended to keep in the range of 0.630 to 0.650 V/cell as shown in Fig. 3 in order to avoid unnecessary over-sulfurization, which tended to create the shunt paths on the absorber surface due to sulfur etching leading to the formation of indium deficient or copper rich areas. To improve the V_{oc}, this would be well controlled and in FY2008, we could achieve the V_{oc} of over 0.680.

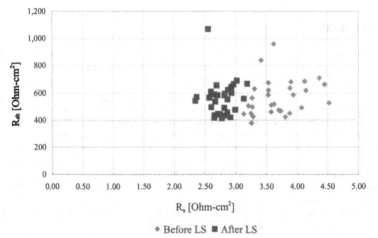

* Before LS ■ After LS

Fig. 4 Correlation of R_{sh} and R_s derived from 30cmx30cm-sized CIS-based thin-film circuits with circuit efficiency over 14 % and measured after 30 min light soaking (LS) under 1 sun condition.

To decrease the R_s, the CBD condition was revised to make a thinner $Zn(O,S,OH)_x$ buffer. Furthermore, the post-deposition annealing in vacuum was applied to force to reduce the OH concentration in the buffer. By applying a thinner $Zn(O,S,OH)_x$ buffer, the composition of this buffer, which is a mixed compound of ZnO, ZnS and $Zn(OH)_2$, might be changed and it would be very likely that compositional change affected the interface quality. Analytical study related to this interface with CBD-$Zn(O,S,OH)_x$ buffer probably thinner than 5 nm is necessary to conclude such experimental assumption. As other approach to increase the R_{sh} and decrease the R_s, thickness and resistivity of MOCVD-ZnO, which was deposited by supplying the diethyl zinc (DEZ) and de-ionized water without any B_2H_6 before depositing the MOCVD-BZO window, were carefully controlled. Making this non-doped ZnO layer functional as a good balancer to prepare better pn interface, double buffer structure with CBD-Zn(O,S,OH)$_x$ buffer and MOCVD-ZnO layers was optimized. Relating to this adjustment, thickness of MOCVD-BZO window was reduced, which contributed to increase the J$_{sc}$. The current level of R$_{sh}$ and R$_s$ derived from 30cmx30cm-sized CIS-based thin-film circuits, which efficiency was over 14 % and measured after 30 min light soaking under one sun condition, is shown in Fig. 4. As seen in this result, R$_s$ shifted to the range of 2.5 to 3.0 after such light soaking. R$_{sh}$ and R$_s$ after light soaking to achieve the FF of over 0.730 was supposed to be 750 and 2.0 on a centered value of the distribution, respectively. This would require better understanding of pn hetero-junction and detailed analytical works on this interface would be necessary to improve the FF. In FY2008, each parameter, V$_{oc}$, J$_{sc}$ and FF, independently achieved the target, such as V$_{oc}$: 0.685 V/cell, FF: 0.735 and J$_{sc}$: 31.8 mA/cm^2, but efficiency was still below 16 %. All of our research works would be focused mainly on the FF in order to achieve the FF of over 0.73 consistently by adjusting the two resistances (R$_{sh}$ and R$_s$) in the 30cmx30cm-sized monolithically integrated circuits.

CONCLUSIONS

The 1st phase on the CIS-based thin-film PV technology in Showa Shell was a technology transfer from the R&D stage to the commercialization or the proof of production capability employing the baseline manufacturing technology developed through the NEDO PV-R&D projects. This phase was recognized to have completed successfully by starting up the 2nd plant of annual production capacity of 60 MW/a in November 2008, in which wide range of operational experience, know-how and lessons of the first plant started up October 2006 were all transferred. Therefore, the 2nd phase on the proof of mass production capability had already started, which should be a real challenge because of a move toward a GW/a production. For realizing this challenge, 16%-efficiency project was started, in which each parameter was targeted, such as V$_{oc}$: 0.685 V/cell, FF: 0.735 and J$_{sc}$: 31.8 mA/cm^2. Up to FY2008, each parameter independently achieved the target, but efficiency was still below 16 %. All of our research works would be focused mainly on the FF in order to achieve the FF of over 0.73 consistently by adjusting the two resistances (R$_{sh}$ and R$_s$) in the circuits. The 16% baseline would be developed by the middle of FY2009 to transfer to the 3rd plant. This achievement would be required to achieve the module (aperture-area) efficiency up to 14 % on a 7200cm^2-sized substrate.

ACKNOWLEDGMENTS

This study was consigned from NEDO.

REFERENCES
[1] K. Kushiya, Tech. Digest of PVSEC17 (2007) 44.
[2] K. Kushiya, M. Ohshita, I. Hara, Y. Tanaka, B. Sang, Y. Nagoya, M. Tachiyuki and O. Yamase. Sol. Energ. Mater. Sol. Cells **75** (2003), p. 171.

[3] D. Abou-Ras, G. Kostorz, D. Bremaud, K. Kälin, F.V. Kurdesau, A.N. Tiwari, M. Döbeli, Thin Solid Films **480-481** (2005) p. 433.
[4] Y. Goushi, H. Hakuma, K. Tabuchi, S. Kijima and K. Kushiya, Tech. Digest of PVSEC17 (2007) 458.

Mater. Res. Soc. Symp. Proc. Vol. 1123 © 2009 Materials Research Society   1123-P05-01

Dielectric Function and Defect Structure of CdTe Implanted by 350-keV Bi Ions

Peter Petrik[1], Miklós Fried[1], Zsolt Zolnai[1], Nguyen Q. Khánh[1], Jian Li[2], Robert W. Collins[2], and Tivadar Lohner[1]
[1]Research Institute for Technical Physics and Materials Science, H-1525 Budapest, POB 49, Hungary
[2]Department of Physics and Astronomy, University of Toledo, Toledo, OH 43606 USA

ABSTRACT

In this work we have developed optical models for the ellipsometric characterization of Bi-implanted CdTe. We have characterized the amount and nature of disorder using Rutherford Backscattering Spectrometry combined with channeling (RBS/C). Samples with a systematically varying degree of disorder were prepared using ion implantation of Bi into single-crystalline CdTe at an energy of 350 keV with increasing doses from 3.75×10^{13} cm^{-2} to 6×10^{14} cm^{-2}. The motivation for use of the high atomic mass Bi ions was that previous studies using lighter ions revealed damage at a low level, even for doses several times higher than the amorphization threshold estimated by simulation [P. Petrik et al., *phys. stat. sol. (c)* 5, 1358 (2008)]. In contrast, Bi ions create sufficient disorder for investigation of the changes in dielectric function critical point (CP) features in a wider variety of structures from single-crystalline to the disordered state. The CP features can be described by numerous methods starting from the standard CP model, through the parameterization of Adachi [Adachi et al., *J. Appl. Phys.* 74, 3435 (1993)], and finally to the generalized CP models. The standard CP model has been demonstrated to be a reliable approach for polycrystalline CdTe characterization used in photovoltaic applications [Li et al., *phys. stat. sol. (a)* 205, 901 (2008)].

INTRODUCTION

This work is a continuation of previous studies on the ion implantation of single-crystalline and polycrystalline CdTe [1]. Our aim is to study the optical properties and the defect structure of disordered CdTe prepared in a controlled way. Ion implantation provides a well established method to introduce defects reproducibly and systematically. Using ion implantation disordered regions of controlled extent can be created, which may be analogous -- in terms of their abilities to scatter optically excited carriers -- to the grain boundary regions in deposited thin film polycrystalline CdTe of different grain sizes. This study supports a better understanding of the optical properties of the thin film CdTe materials used for solar applications [2] (For example, using a low-temperature sputter-deposition technology for CdS/CdTe, a solar cell efficiency of 14 % has been demonstrated [3].)

EXPERIMENT

c-CdTe (Nippon Mining & Metals Co., (111)B) samples have been implanted using 175 kV Bi^{2+} ions at fluences ranging from 3.75×10^{13} cm^{-2} to 6×10^{14} cm^{-2}. The high-mass Bi ions were chosen because the damage created in previous studies using Xe implantation was too low

[1]. Using double-charged Bi ions at 175 kV, each Bi ion will have an energy of 350 keV. Simulation with the SRIM (Stopping and Range of Ions in Matter, [6]) software suggests a highly damaged region within the top 175 nm of the sample and a peak position near the surface (about 50 nm, see Fig. 1). We note that the SRIM simulation was performed assuming a target density of 6.2 g/cm^3 for CdTe.

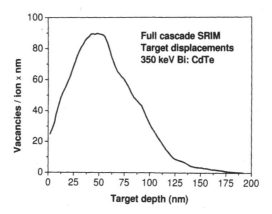

Figure 1. SRIM simulation of disorder for 350 keV Bi ions implanted into CdTe.

The ellipsometry spectra (ψ, Δ) of the virgin and ion-implanted samples were measured over the spectral range of 1 - 6 eV (covering the critical point energies of E_0 [~1.5 eV], E_1 [~3.3 eV], $E_1+\Delta_1$ [~3.9 eV], and E_2 [~5.1 eV]) using a rotating-compensator multichannel ellipsometer [7] similar to that developed to study Si:H-based solar cells [4, 5]. The pseudo-dielectric function spectra reveal significant damage through the broadening of the critical point (CP) features, although full amorphization is not attainable (see Fig. 2).

The depth distribution of disordered atoms was simulated by the SRIM software [6] and measured by Rutherford Backscattering Spectrometry combined with Channeling (RBS/C). The backscattered He$^+$ ions were detected using an ORTEC surface barrier detector mounted in Cornell geometry at a scattering angle of 165°. The RBS/C spectra were evaluated using the RBX software [8]. The RBS/C spectra reveal only a dechanneling contribution caused by extended defects (e.g. dislocation loops; see Fig. 3) and not by point defects. The lack of a peak and the appearance of spectra well below the random level indicate low-level damage without amorphization, as also shown by the Fig. 2 pseudo-dielectric function spectra, determined from (ψ,Δ) using the Fresnel equations for a single ambient/CdTe interface. The RBS/C spectra show that the defects give mainly dechanneling contribution and not direct backscattering, since there is no decrease in the backscattering yield after the maximum range of the Bi ions (In Fig. 3 between channels 175-180, corresponding to a depth of about 180 nm in CdTe).

82

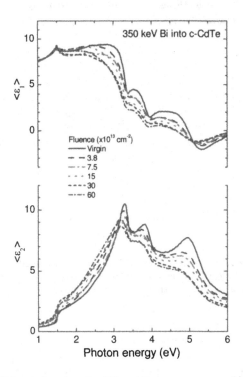

Figure 2. Pseudo-dielectric functions for different fluences as measured by ellipsometry.

DISCUSSION

As Fig. 1 shows most of the Bi ions are stopped in the uppermost 100 nm of CdTe. Since the peak Bi concentration remains below 0.3 atom % even for the highest fluence the contribution of Bi to the RBS/C spectra is negligble and there is no significant background from Bi to the CdTe signal, see Fig. 3.

The RBS/C spectra can be analyzed to estimate the relative concentration of defects as a function of the fluence. The slope of the RBS/C spectra in the damaged region and the height of the spectra near the end of range of defects (shown by vertical lines in Fig. 3) can be used for this purpose. Assuming an average dechanneling contribution of defects in the implanted zone the dechanneling probability for He projectiles for unit depth is calculated with the RBX program. As the results show in Fig. 4 for a fluence of e.g. 3×10^{14}/cm^2 the average dechanneling probability per unit depth is about 2×10^4/cm. In this case the probability for an initially channeled He ion for dechanneling upon crossing the damaged zone with a thickness of 180 nm is about 0.4. This ratio is reflected also in Fig. 3 in the heights of the implanted and random spectra within the marked window. The results in Fig. 4 suggest a nearly linear behavior for the amount of defects vs the implantation fluence.

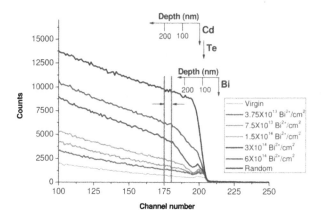

Figure 3. 1.5 MeV He⁺ RBS/C spectra of CdTe implanted with different Bi^{2+} fluences. The depth scales for Bi and Cd/Te are indicated (Cd and Te have nearly the same atomic mass). The window between channels 175-180 shows the region of the end of range of defects given by the SRIM program.

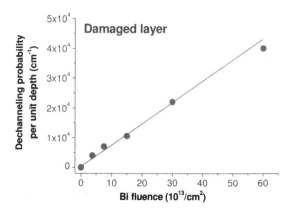

Figure 4. Average dechanneling probability in the damaged zone of implanted CdTe calculated from the RBS/C spectra using the RBX [8] spectrum simulation code.

84

The dielectric function of the damaged region is calculated using an overlayer correction. The overlayer was taken into account as a surface roughness layer, consisting of a mixture of material and voids. The material component is assumed to exhibit optical properties of the Cauchy form. We used a multi-sample method for fitting the overlayer thickness, the volume fraction of voids in the overlayer, the Cauchy parameters, and the real and imaginary parts of the dielectric function of the CdTe, using as data the (ψ,Δ) spectra at three angles of incidence. In this method the spectra were fit point-by-point for all samples simultaneously. The Cauchy expression used here for the complex refractive index included three terms for the real part and two for the imaginary part. The resulting five parameters were assumed to be independent of sample (Fig. 5), with the sample dependence in the overlayer accounted for by variations in void volume fraction and thickness.

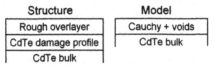

Figure 5. Optical model used to determine the dielectric function of disordered CdTe.

The CP features can be quantified by numerous methods of increasing complexity starting from the standard critical point model [7], through Adachi's parameterization [10], to the generalized CP models [11]. The dielectric functions of ion implanted CdTe were fitted at each CP using a standard analytic line shape

$$\varepsilon(E) = Ae^{i\varphi}(E_{CP} - E - i\Gamma)^{\mu}, \tag{1}$$

where A, E_{CP}, Γ, and φ are the amplitude, the CP energy, the broadening, and the excitonic phase angle, respectively. The fitted energies and broadening parameters are plotted as functions of the fluence in Fig. 6 for the CPs of E_1, $E_1+\Delta_1$, and E_2. The CP energies E_1 and $E_1+\Delta_1$ show a systematic decrease with increasing fluence; however, the apparent increase of E_2, which is most strongly influenced by the surface, is not significantly outside the error bars . The broadening parameters (Γ) of all CPs reveal an increase with increasing fluence, as expected. The drop of $\Gamma(E_2)$ at the highest fluence may be due to a near-surface recrystallization, which may also lead to the observed complexity in the E_2 energy. The thickness of the rough overlayer increases from about 8 to 20 nm with increasing dose, whereas the volume fraction of voids in the rough overlayer increases from about 20 to 45 %, as expected. The latter is in line with the assumption that the impinging ions create craters of increasing surface density with increasing fluence.

CONCLUSIONS

In this work we have characterized 350-keV Bi ion implanted single-crystalline CdTe using spectroscopic ellipsometry and RBS/C. We have obtained a degree of disorder that is much greater than that for Xe ion implantation investigated earlier. The RBS/C results show the presence of extended defects, the concentration of which linearly increases as a function of increasing fluence. We used a multi-sample approach together with rough overlayer correction to determine the dielectric function of the disordered CdTe. The dielectric function was analyzed using a standard analytical lineshape. In agreement with RBS/C, the ellipsometric results reveal a systematic change of broadening and shift of the E_1 and $E_1+\Delta_1$ CP energies to lower values.

ACKNOWLEDGMENTS

Support from the Hungarian Scientific Research Fund (OTKA Nr. K61725) is greatly acknowledged.

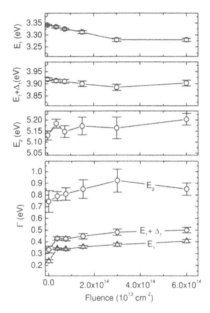

Figure 6. Fitted CP energies (E) and broadening parameters (Γ) of Eqn. 1 as a function of the fluence.

REFERENCES

1. P. Petrik, N. Q. Khánh, Jian Li, Jie Chen, R. W. Collins, M. Fried, G. Z. Radnóczi, T. Lohner, and J. Gyulai, phys. stat. sol. (c) 5, No. 5, 1358–1361 (2008).
2. Jian Li, Jie Chen, N. J. Podraza, and R. W. Collins, Proceedings of the 4th World Conference on Photovoltaic Energy Conversion, May 2006, Waikoloa, Hawaii, USA, IEEE Proceedings ISBN 1-4244-0017-1, Volume 1, 392 (2006).
3. A. Gupta and A. D. Compaan, Appl. Phys. Lett. 85, 684 (2004).
4. Joungchel Lee, P. I. Rovira, Ilsin An, and R. W. Collins, Rev. Sci. Instr. 69, 1800 (1998).

5. R. W. Collins, Ilsin An, Joungchel Lee, and J. A. Zapien, in Handbook of Ellipsometry (William Andrew, Norwich, NY, 2005), p. 481.
6. The Stopping and Range of Ions in Matter, http://www.srim.org.
7. B. Johs, J.A. Woollam, C.M. Herzinger, J.N. Hilfiker, R. Synowicki, and C. Bungay, *Proc. Soc. Photo-Opt. Instrum. Eng., Crit. Rev. 72, 29 (1999).*
8. E. Kótai, Nucl. Instr. and Meth. B 85, 588 (1994).
9. P. Lautenschlager et al., Phys. Rev. B 36, 4821 (1987).
10. S. Adachi et al., J. Appl. Phys. 74, 3435 (1993).
11. B. Johs et al., Thin Solid Films 313-314, 137 (1998).

Mater. Res. Soc. Symp. Proc. Vol. 1123 © 2009 Materials Research Society   1123-P07-02

MOCVD Growth of High-Hole Concentration (>2×10^{19} cm^{-3}) P-Type InGaN for Solar Cell Application

Hongbo Yu[1], Andrew Melton[1], Omkar Jani[2], Balakrishnam Jampana[3], Shenjie Wang[1], Shalini Gupta[1], John Buchanan[1], William Fenwick[1], and Ian Ferguson[1]

[1] School of Electrical and Computer Engineering, Georgia Institute of Technology, Atlanta, GA 30332

[2] Institute of Energy Conversion, University of Delaware, Newark, DE 19716
[3] Materials Science and Engineering Department, University of Delaware, Newark, DE 19716

ABSTRACT

InGaN alloys are widely researched in diverse optoelectronic applications. This material has also been demonstrated as a photovoltaic material. This paper presents the study to achieve optimum electrically active p-type InGaN epi-layers. Mg doped InGaN films with 20% In composition are grown on GaN templates/sapphire substrates by MOCVD. It is found that the hole concentration of p-type InGaN depends strongly on the Mg flow rate and V/III molar ratio and hole concentration greater than 2×10^{19} cm^{-3} has been achieved at room temperature. The optimum activation temperature of Mg-doped InGaN layer has been found to be 550-600°C, which is lower than that of Mg-doped GaN. A solar cell was realized successfully using the InGaN epi-layers presented here.

INTRODUCTION

InGaN alloys have achieved commercial success in III-Nitride based light emitting diodes (LEDs) and laser diodes (LDs). InGaN alloys have a wide bandgap range from 0.7 eV (InN) to 3.4 eV (GaN) covering most of the solar spectrum. This wide span of InGaN band gap makes InGaN an ideal material system for realization of high efficiency solar cells [1].

Achieving electrically active p-type InGaN is essential for realizing III-Nitride photovoltaic devices. Although there are many reports on the structural and optical properties of InGaN films, there are few studies on the metal organic chemical vapor deposition (MOCVD) growth and p-type electrical properties [2, 3]. Nitrogen vacancies created by low growth temperatures cause difficulty in p-type doping of InGaN. In this paper the MOCVD growth and electrical studies of p-type InGaN layers with 20% indium content are presented. Also presented are device results achieved by incorporating these epitaxial layers into a solar cell structure.

EXPERIMENT

Samples used in this study were grown on c-plane (0001) sapphire substrates by low pressure MOCVD. First a 2μm thick undoped GaN film was grown at 1050°C on a 25nm GaN nucleation layer deposited at 500°C. Next Mg-doped InGaN layer was grown on the undoped GaN film using N_2 as carrier gas. Trimethylindium (TMIn), Triethylgallium (TEGa), Bis-magnesium (Cp2Mg), and ammonia (NH_3) were used as the precursors for In, Ga, Mg and N, respectively. The thickness of Mg-doped InGaN was around 100nm in all samples.

The indium content in the InGaN alloys was calibrated by changing the growth temperature. After growth, the samples were activated using rapid thermal annealing in N_2 ambient. The electrical properties of p-type InGaN were studied by Hall effect measurement using the standard Van Der Pauw technique.

RESULTS and DISCUSSION

The indium content in the InGaN layers depend strongly on the growth temperature, which is reduced from 30% to 15% when the growth temperature increases from 700 to 760°C. Figure 1 shows the XRD patterns of the samples grown at different temperatures with other growth conditions kept constant. As shown in this figure, as the indium fraction increases the intensity of the InGaN XRD peak decreases and the full width at half maximum (FWHM) increases. This implies deterioration in crystal quality of the InGaN layer with increasing indium content in the InGaN alloys. Here we choose an InGaN alloy with indium content of approximately 20% to correlate the electrical properties of the p-type InGaN with MOCVD growth conditions. The V/III molar ratio used in the growth of these samples was 39300.

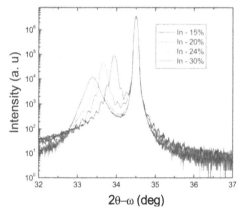

Figure 1. XRD patterns of InGaN layers grown on GaN templates using different growth temperatures

Electrically active p-type GaN is achieved by removing the atomic hydrogen from the acceptor-H neutral complex formed during growth. This is realized either by using low-energy electron-beam irradiation (LEEBI) treatment or thermal annealing at above 700°C [4, 5]. Thus far there is no report on the effect of thermal annealing temperature on the electrical properties of Mg-doped InGaN alloy. A previous report on p-type InGaN used 700°C thermal annealing, which is consistent with the thermal annealing process of p-type GaN [3, 6]. However, the growth temperature of InGaN was much lower than GaN growth, and InGaN alloys are unstable at high temperature. Based on this it is suspected that the optimum thermal annealing temperature for InGaN should be lower than that for GaN.

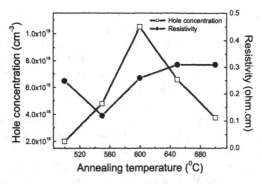

Figure 2. Electrical characteristics as a function of the annealing temperature obtained for Mg-doped InGaN

Figure 2 shows the hole concentration and resistivity of p-type $In_{0.2}Ga_{0.8}N$ as a function of annealing temperature. The thermal annealing temperature was varied from 500 to 700°C, and the annealing time was set to 15 min. As shown in Figure 2 the hole concentration and resistivity of p-type InGaN layers display a strong dependence on the thermal annealing temperature. The lowest resistivity and highest hole concentration are obtained at 550°C and 600°C, respectively. When the temperature was increased to 700°C the hole concentration is reduced to about half of the value seen at 600°C. Our experimental results suggest that the suitable thermal annealing temperature for Mg-doped $In_{0.2}Ga_{0.8}N$ was 550-600°C, which is approximately 100°C lower than the reported annealing temperature for Mg-doped GaN. It is believed that the reason for the increased hole concentration with decreased annealing temperature may be the suppression of nitrogen atom disassociating from the InGaN alloy, which occurs at higher temperatures. The n-type characteristics observed in samples annealed at high temperature could be due to nitrogen vacancies acting as n-type donors in InGaN alloys. Phase separation of InGaN alloys with indium content higher than 10% has been previously reported in annealing experiments at 600 and 700°C, which may also influence the hole concentration in InGaN [7].

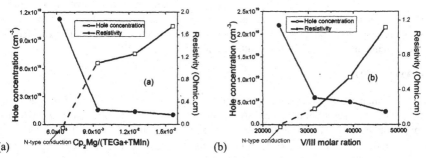

Figure 3. Variations of hole concentration and resistivity with $Cp_2Mg/(TEGa+TMIn)$ ratio (a) and V/III molar ratio (b) for p-type InGaN annealed at 600°C

The influence of growth conditions on p-type conductivity was investigated. Figure 3a and 3b show the variations of hole concentration and resistivity with $Cp_2Mg/(TEGa+TMIn)$ ratio and V/III molar ratio, respectively. For this study the annealing temperature was 600°C for all samples. As shown in Figure 3a, the Mg-doped InGaN layers changed from n-type to p-type conduction after annealing, with hole concentration of 1.05×10^{19} cm^{-3} when the $Cp_2Mg/(TEGa+TMIn)$ ratio was increased from 6.25×10^{-3} to 1.56×10^{-2}. To achieve p-type conduction the Mg incorporation must exceed the n-type background carrier concentration of undoped InGaN. In addition, it has been found that the electrical characteristics of Mg-doped InGaN layers depend strongly on the V/III molar ratio during InGaN growth, as shown in Figure 3b. As the V/III ratio was increased from 2.36×10^{5} to 4.72×10^{5} the Mg-doped InGaN layer turned from n-type to p-type conduction with hole concentration of 2.15×10^{19} cm^{-3}. It is believed that a relatively high V/III ratio is needed to improve the p-type conductivity of Mg-doped InGaN layer, possibly to suppress nitrogen vacancy formation during growth. Under optimized growth and annealing conditions the lowest resistivity of p-type InGaN film was measured to be 0.1 Ω-cm, which is about 10 times lower than that of typical p-type GaN film.

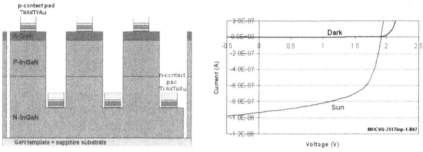

Figure 4. Device schematic and I-V curves of the test solar cell using optimized high hole concentration InGaN as p-type layer

Based on the high hole concentration InGaN discussed here, single junction solar cells have been successfully demonstrated. The device structure were grown on undoped GaN and constitute a 200nm Si doped n-type InGaN layer, 100nm Mg doped p-type InGaN and a strained 10nm n-type GaN capping layer. Figure 4a and 4b show the schematic device structure and corresponding I-V curve of the test solar cell. The cells exhibited an open-circuit voltage (V_{oc}) of 1.95 V with a fill factor (FF) of 57.3 %.

CONCLUSION

In summary, the influence of growth conditions and thermal annealing temperature on the electrical properties of p-type InGaN was investigated. The optimum annealing conditions for p-type $In_{0.2}Ga_{0.8}N$ are observed to lie between 550°C and 600°C. P-type InGaN with hole concentration greater than 2×10^{19} cm^{-3} at room temperature was achieved. The lowest resistivity of p-type InGaN film was measured to be 0.1 Ω-cm, which is about 10 times lower than that of normal p-type GaN film. A test InGaN based solar cell was realized successfully using the optimized high hole concentration and low resistive p-type InGaN.

REFERENCES

[1] O. Jani, I. Ferguson, C. Honsberg, S. Kurtz, Applied Physics Letters 91 (2007) 132117.

[2] S. Yamasaki, S. Asami, N. Shibata, M. Koike, K. Manabe, T. Tanaka, H. Amano, I. Akasaki, Applied Physics Letters 66 (1995) 1112-1113.

[3] K. Kumakura, T. Makimoto, N. Kobayashi, Japanese Journal of Applied Physics, Part 2 (Letters) 39 (2000) 337-339.

[4] H. Amano, M. Kito, K. Hiramatsu, I. Akasaki, Japanese Journal of Applied Physics, Part 2 (Letters) 28 (1989) 2112-2114.

[5] S. Nakamura, N. Iwasa, M. Senoh, T. Mukai, Japanese Journal of Applied Physics, Part 1 (Regular Papers & Short Notes) 31 (1992) 1258-1266.

[6] S. Nakamura, M. Senoh, T. Mukai, Japanese Journal of Applied Physics, Part 2 (Letters) 30 (1991) 1708-1711.

[7] K. Osamura, S. Naka, Y. Murakami, Journal of Applied Physics 46 (1975) 3432-3437.

Mater. Res. Soc. Symp. Proc. Vol. 1123 © 2009 Materials Research Society   1123-P06-01-F07-01

FASST® Reactive Transfer Printing for Morphology and Structural Control of Liquid Precursor Based Inorganic Reactants

P. Hersh[1], M. Taylor[1], B. Sang[1], M. van Hest[2], J. Nekuda[2], A. Miedaner[2], C. Curtis[2], J. Leisch[2], D. Ginley[2], B.J. Stanbery[1], and L. Eldada[1]

[1]HelioVolt Corporation, 8201 E. Riverside Dr., Austin, TX 78744, U.S.A.
[2]National Renewable Energy Laboratory, Golden, CO 80401, U.S.A.

ABSTRACT

Field-Assisted Simultaneous Synthesis and Transfer (*FASST®*) process offers a controllable and cost-effective method to produce Copper Indium Gallium Selenide (CIGS) films for high efficiency photovoltaic devices. In the first stage of the two-stage *FASST®* process two separate precursor films are formed, one deposited on the substrate and the other on a reusable printing plate. In the second stage, the precursors are brought into intimate contact and rapidly reacted under pressure in the presence of an applied electrostatic field, effectively creating a sealed micro-reactor that ensures high material utilization efficiency, direct control of reaction pressure, and low thermal budget. The unique ability to control both precursor films independently allows for composition and deposition technique optimization, eliminating pre-reaction prior to the synthesis of CIGS. This flexibility has proven immensely valuable as is demonstrated in the results of depositing the two-reactant films by various combinations of low-cost solution-based and conventional vacuum-based physical vapor deposition techniques, producing in several minutes' high quality "hybrid" CIGS with large grains on the order of several microns. Cell efficiencies as high as 12.2% have been achieved using the *FASST®* method.

INTRODUCTION

Copper Indium Gallium Selenide (CIGS) photovoltaic cells have demonstrated efficiencies up to 19.9% for small-area laboratory cells [1] and 13% for large-area modules [2]. However, manufacturability has been a challenge for CIGS, and high throughput, low-cost methods of producing high quality devices have yet to be fully realized. This paper will describe HelioVolt's novel, rapid and cost-effective CIGS manufacturing process, Field-Assisted Simultaneous Synthesis and Transfer (*FASST®*) [3], and its use of solution-deposited precursors [4]. The *FASST®* process is being developed to manufacture high efficiency CIGS modules and is currently being scaled up on a 20 MW production line.

EXPERMENTAL DETAILS

High-performance CIGS is characterized by relatively large grain sizes and the formation within individual grains of a nanoscale interpenetrating network of copper-rich and copper-deficient domains which form a percolation network for electrons and holes respectively [5,6]. The conventional method used to synthesize high performance thin film

CIGS devices is a high-temperature co-evaporation method. The multi-step deposition sequences developed to achieve this performance always involve the topotactic transformation of a large-grain precursor into CIGS rather than the direct synthesis of CIGS from condensation of elemental vapors as in molecular beam deposition. This same topotactic transformation is used by HelioVolt's proprietary FASST® process.

The *FASST*® process utilizes a two-stage reactive transfer printing method relying on chemical reaction between two separate precursor films to form CIGS. A schematic of the *FASST*® process is shown in Fig. 1. In the first stage, two Cu-In-Ga-Se-based precursor layers, forming the chemical basis of CIGS, are deposited onto a substrate and a print plate, respectively. The two separate precursors provide the benefit of independently optimized composition, structure, deposition method, and processing conditions for each precursor. Separating the precursors eliminates pre-reaction prior to the second stage *FASST*®, and facilitates optimized CIGS formation in the second stage. Furthermore, precursors can be deposited at a low substrate temperature enabling lower cost, and higher throughput.

In the second stage, these precursors are brought into intimate contact and rapidly reacted under the pressure in the presence of an applied electrostatic field. The method utilizes physical mechanisms characteristic of rapid thermal processing (RTP) and anodic wafer bonding (AWB), effectively creating a sealed micro-reactor that insures high material utilization, direct control of reaction pressure, and low thermal budget. The rapid thermal transient provides the similarity between the *FASST*® process and RTP. By pulse heating the film through the print plate, the overall thermal budget is significantly reduced, allowing the use of low cost less thermally stable substrate materials.

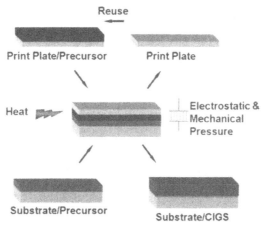

Figure 1: Schematic of the *FASST*® process.

Sufficient mechanical pressure can substantially prevent the loss of Se vapor from the reaction zone, thereby achieving highly efficient incorporation of Se into the composition layer. The use of an electrical bias between the print plate and substrate creates between them an attractive force that serves to insure intimate contact between the precursor films on an atomic scale, and can thus be used in conjunction with mechanical pressure to control the total pressure in the reaction zone. This is the resemblance between the *FASST*® process and AWB, a method developed historically to reduce the temperature required to bond two dissimilar materials together.

DISCUSSION

CIGS by *FASST*®

Large grain high quality CIGS is synthesized from two precursors in several minutes using the *FASST*® process. Figure 2 shows a secondary ion mass spectrometry (SIMS) depth profile of a CIGS thin film processed by the *FASST*® in only six minutes. The precursors for the film are made by physical vapor deposition (PVD) methods. The uniform elemental distribution indicates a complete reaction of the precursors, and the X-ray diffraction (XRD) analysis confirms the absence of deleterious phases other than CIGS.

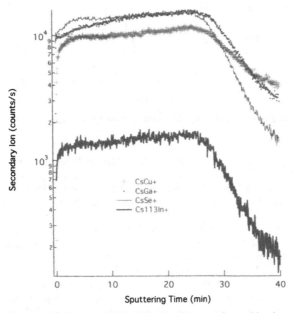

Figure 2: SIMS depth profile of a CIGS film. This film was formed in six minutes by the *FASST*® process.

Figure 3 shows the XRD pattern. All the peaks are indexed based on chalcopyrite-type CIGS and Mo structure. The *FASST*® processed film has a (220/204) preferred orientation. Evidence indicated that the (220/204) oriented films may help junction formation and improve solar cells performance [7]. The rapid processing of CIGS formation significantly increases the manufacturing throughput. As described above, this unique processing approach results in a much lower thermal budget as compared to co-evaporation and two-step selenization processes, which are common CIGS manufacturing. The lower thermal budget, removal of selenization process and high throughput all contribute to a low cost process leading to improved manufacturability.

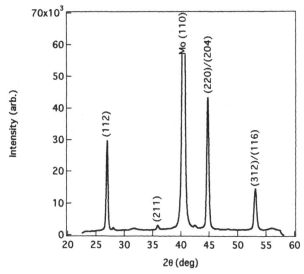

Figure 3: XRD pattern of a CIGS film fabricated by the *FASST*® process.

The opportunity to tailor the two precursors independently allows for the use of unconventional, non-vacuum deposition techniques such as ultrasonic spraying, dye coating, inkjet printing, ultrasonic spraying, direct writing, and screen printing. These atmospheric-pressure-based deposition tools offer great flexibility and open up entirely new windows for materials processing. They also offer a viable means of introducing nanoparticle technology, metal organic chemistry and novel reaction paths to produce CIGS. The low capital equipment cost and high throughput capabilities associated with atmospheric pressure processing potentially reduce manufacturing cost. These materials can be deposited in air at temperatures below 200°C, which will lower the thermal budget [8].

Proprietary inks containing a variety of soluble Cu-, In- and Ga- binary selenide materials have been developed. These metal-organic inks are designed to decompose into the desired precursors. The resultant precursors are then used in step one of the *FASST*® process.

For the work described in this manuscript, the inks were deposited using an ultrasonic spray head fed by a variable speed liquid pump. A substrate heater mounted on a computer-controlled X-Y motion system allowed for movement of heated substrates under the sprayed stream. The thickness of the sprayed film was controlled by varying the ink concentration, the flow rate through the sprayer and the number of coats sprayed. Conditions were optimized such that smooth, uniform precursor films were obtained for all of the sprayed inks.

The precursor films were converted to the desired materials through rapid thermal processing (RTP) in a controlled atmosphere. The RTP conditions were varied systematically to ascertain the effect of conditions on the film compositions and morphologies obtained. The film compositions were characterized by X-ray fluorescence (XRF), crystalline phases were identified using XRD and film morphology was examined using scanning electron microscopy (SEM).

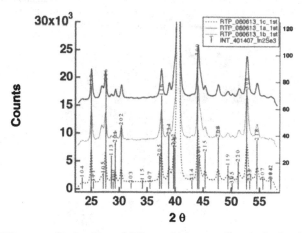

Figure 4: XRD scans of In₂Se₃ films of different thicknesses deposited using the In-containing ink. The large peak at ~40° is due to Mo.

In this work, inks developed for deposition of binary selenide materials were deposited as one of the precursors, and a PVD-deposited thin film was used as the other precursor. In one case, an In-containing precursor film was sprayed on a heated, Mo-coated glass substrate and processed by RTP to give In₂Se₃. Figure 4 shows XRD scans of thickened In₂Se₃ films processed under these conditions.

After the RTP step these liquid precursor films were found to be cracked and somewhat porous. However when they were sprayed at a low temperature then processed with a slower temperature ramp to a higher temperature, a dense film was obtained, as shown in the SEM micrograph of Figure 5. This result represents a breakthrough for solution-based precursors. All temperatures used were below 200°C.

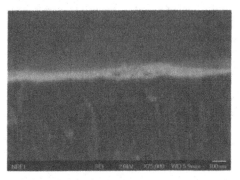

Figure 5: SEM image of a precursor film produced from an ink sprayed at low temperature and processed with a slow temperature ramp.

CIGS films were manufactured by the *FASST®* process using a combination of PVD-based and ultra-sonic spray deposited precursors. Figure 6 shows cross-sectional and top view SEM micrographs of such a film, revealing high quality large columnar grains up to 4 µm in size.

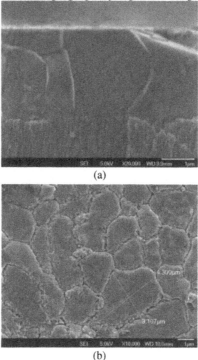

(a)

(b)

Figure 6: SEM micrographs of (a) the cross section and (b) the top view of a CIGS film synthesized using a non-vacuum-deposited precursor.

The XRD pattern for the Figure 6 sample is shown in Figure 7 and the chalcopyrite CIGS phase is clearly identified. Again, (220/204) textured film is made by the *FASST®* process. The CIGS films are being applied to solar devices.

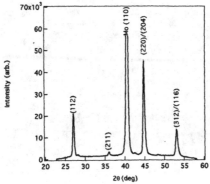

Figure 7: XRD pattern of a CIGS film by the *FASST®* process using a non-vacuum deposited precursor.

Another benefit of the *FASST®* process is that there is no constraint on the combination or type of precursors that can be brought together. The only requirement is that all of the elements in the correct stoichiometry must be present on the substrate and printing plate prior to the *FASST®*.

From a manufacturing standpoint, any deposition method, whether it is PVD-based or atmospheric- pressure-based, has relative advantages and disadvantages. While the capital equipment cost for atmospheric-pressure-based systems is lower than that of the corresponding PVD systems, the raw material costs tend to be higher for solution rather than for PVD sources. Therefore, the best choice is ultimately governed by differences in the performance and yield of products manufactured by these two approaches. Better material utilization coupled with the decreasing cost of liquid precursors due to the maturation of the nanotechnology field make atmospheric- pressure-based processing a more attractive alternative.

Solar Device Results

Solar cells with a conventional device structure of glass/Mo/CIGS/buffer/TCO were fabricated. The CIGS absorber in these cells was formed by the *FASST®* process with PVD-based precursors. The SEM micrograph in Figure 8 depicts a cross section of a representative device. As can be seen, high-quality CIGS films with large columnar grains are obtained.

101

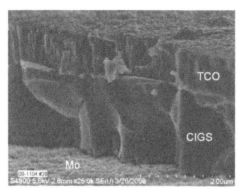

Figure 8: SEM cross-sectional image of a CIGS device.

External quantum efficiency (QE) versus wavelength for the device without anti-reflection (AR) coating is shown in Fig. 9. High QE at wavelengths over 550 nm reveals very good carrier collection and good performance of the CIGS layer. A low QE at short wavelengths indicates the need to further optimize window layers.

Solar cells of over 12% efficiency have been fabricated using the *FASST*® process. An I-V curve of a 12.2% efficient device is shown in Fig. 10. An anti-reflection layer was deposited on this device and the efficiency was confirmed by Colorado State University. The composition of the CIGS film in this device as measured by XRF is 21.8% Cu, 21.6% In, 6.3% Ga and 50.4% Se, which gives a Cu ratio (Cu/(In+Ga)) of 0.78 and Ga ratio (Ga/(In+Ga)) of 0.22. Optimizing Cu and Ga ratios should increase device efficiency by the *FASST*® process. For instance, an open circuit voltage of 590 mV was obtained by increasing the Ga/(In+Ga) ratio to 0.3.

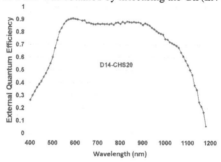

Figure 9: QE curve of a CIGS solar cell.

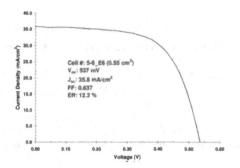

Figure 10: J-V curve of a CIGS solar cell measured by Colorado State University.

Analysis of the J-V data of the device gave a diode quality factor of about 2, and high saturation current density, which means that the large recombination at the junction region limits the open circuit voltage and fill factor.

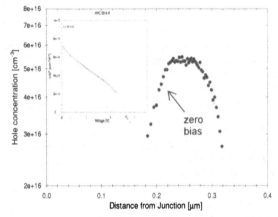

Figure 11: Carrier density as a function of distance from the junction, as derived from a C-V measurement.

Further analysis of the device was carried out by capacitance-voltage (C-V) measurement [9]. This measurement gave a hole density of 2.5×10^{16} cm^{-3} and a depletion width of ~0.2 μm. Carrier concentration as a function of distance from the junction, as derived from C-V data, is shown in Fig. 11. The hump in carrier density against distance might be a signature of the measurement responding to deep states near the interface [9,10], and direct measurement of deep-level defects would be needed to verify it. Such states have a detrimental effect on the solar cell efficiency, because they constitute effective recombination paths for forward current opposing the photo-generated current. Such forward currents might also result from enhanced tunneling recombination through these states [11]. This suggests that the interface region, including the CIGS surface termination and post-CIGS treatment, need to be further optimized. Elimination of these states by improving the CIGS surface termination could significantly improve the open-circuit voltage and the fill factor.

103

CONCLUSIONS

The Field-Assisted Simultaneous Synthesis and Transfer ($FASST^®$) process was described. The $FASST^®$ process is a two-stage reactive transfer printing method relying on chemical reaction between two separate precursor films to form CIGS. The very short processing time, low thermal budget, and high material utilization, lead to a low cost process. (220/204) textured CIGS films with large grains on the order of several microns were made in just several minutes by the $FASST^®$ process, using both PVD-based precursors and non-vacuum- deposited precursors. Initial test devices showed efficiencies of over 12%. HelioVolt's proprietary $FASST^®$ process is currently being scaled up on a 20 MW production line.

ACKNOWLEDGEMENTS

The authors thank Tim Nagle, Alan Davies and Jim Sites of Colorado State University for the confirmation I-V and C-V measurements and helpful discussions.

REFERENCES

1. I. Repins, M. Contreras, B. Egaas, C. DeHart, J. Scharf, C. Perkins, B. To, and R. Noufi, *Proc. Photovolt. Res. Appl.* **13**, 209 (2008).
2. M. Powalla, *Proc. Euro. Photovolt. Solar Energy Conf.* **21**, 1789 (2006).
3. B. J. Stanbery, US Patent 6,881,647.
4. L. Eldada, F. Adurodija, B. Sang, M. Taylor, A. Lim, J. Taylor, Y. Chang, S. McWilliams, R. Oswald, and B.J. Stanbery, *Proc. Euro. Photovolt. Solar Energy Conf.* **23**, 2142 (2008).
5. B. J. Stanbery, *Critical Reviews in Solid State and Materials Sciences* **27**, 73 (2002).
6. Y. Yan, R. Noufi, K. M. Jones, K. Ramanathan, M. M. Al-Jassim, and B. J. Stanbery, *Applied Physics Letters* **87**, 121904 (2005).
7. S. Chaisitsak, A. Yamada, and M. Konagai, *Jpn J Appl Phys.* **41**, 507 (2002).
8. C. Curtis, M. Hest, A. Miedaner, J. Nekuda, P. Hersh, J. Leisch, and D. Ginley, *Proc. IEEE Photovolt. Special. Conf.* **33**, 1065 (2008).
9. P. Mauk, H. Tavakolian, and J. Sites, *IEEE Trans. on Electron Devices* **37**, 422 (1990).
10. H. Tavakolian and J. Sites, *Proc. IEEE Photovolt. Special. Conf.* **20**, 1065 (1988).
11. I. Repins et al., *Proc. Photovolt. Res. Appl.* **14**, 25 (2006).

Mater. Res. Soc. Symp. Proc. Vol. 1123 © 2009 Materials Research Society   1123-P05-06

Electrodeposition of Cu$_2$ZnSnS$_4$ Thin Films Using Ionic Liquids

C.P.Chan[1], H. Lam[1], K.Y. Wong[2] and C. Surya[1,*]

[1]Department of Electronic and Information Engineering and Photonics Research Centre
The Hong Kong Polytechnic University, Hong Kong, China

[2]Department of Applied Biology and Chemical Technology,
The Hong Kong Polytechnic University, Hong Kong, China

* Corresponding author

ABSTRACT

We report the growth of Cu$_2$ZnSnS$_4$ (CZTS) thin films by electrodeposition in ionic liquid. Sulfurization was performed in elementary sulfur vapor environment at 450°C for 2 hours. The X-ray diffraction analysis indicated that the film has a stannite structure with preferred grain orientation along (112). Photo-absorption measurement of the sample was performed from 500 nm to 990 nm. It is found that the energy bandgap of the film is about 1.49eV and the absorption coefficient is found to be of the order of 10^4cm^{-1}.

INTRODUCTION

It is highly desirable to fabricate high conversion efficiency solar materials at low cost. Among the thin film photovoltaic materials, copper-based chalcopyrite, copper indium gallium selenide (CIGSe) absorbers have been intensively studied in the past decade [1]. Laboratory CIGSe solar cells show high conversion efficiency of 19.9% [2] and high optical absorption coefficient. While CIGSe-based photovoltaic cells demonstrate desirable optoelectronic properties for the development of high-efficiency cells, the mass production of devices faces difficulties on the scarcity of indium and toxicity of selenium, leading to a concern for its viability for large-scale deployment.

Recently, Copper-Zinc-Tin-Sulfide (CZTS), has attracted growing attention as a potential photovoltaic material due to the high absorption coefficient of the film, the abundance and the non-toxicity of the constituent elements, making large-scale deployment of the device a feasibility.

Vacuum deposition methods by electron beam and sputtering techniques have been widely investigated for the preparation of CZTS and other photovoltaic materials, but these techniques involve high capital investment, thus only applicable to high value niche markets. For this reason, electrodeposition of semiconductors has emerged as an alternative method for preparing thin film solar cell materials. The advantages of the electrodeposition technique are: i.) the process does not involve expensive vacuum modules; ii.) the technique provides excellent utilization of the raw materials; and iii.) the possibility of large area deposition of the material. The success development of the technique may lead to significant lowering in the cost of the materials.

The use of ionic liquids as the electrolyte for the deposition process offers some promising benefits over the commonly used aqueous solutions [3]. The ionic liquid used in this experiment are both air and water stable, which has negligible vapor pressure up to 130°C, allowing film deposition at high temperatures. Moreover, the electrolyte has an electrochemical potential window of 2.5 V (-1.25V to +1.25V *vs.* Ag) and has the high conductivity required for electrochemical applications. This property allows the reduction of highly reactive metals such as aluminium and zinc at a better quality, of which the electrodepositions in aqueous solutions are difficult owing to a massive hydrogen evolution at the working electrode, leading to hydrogen embitterment.

EXPERIMENT DETAILS

In the present study, we report on the electrodeposition of CZTS thin films from metal salts using choline chloride based ionic liquid. The ionic liquid was prepared by mixing one mole of choline chloride ($C_5H_{14}ONCl$, from Aldrich) to two moles of urea ((NH_2)2CO, from Aldrich). The mixtures were obtained by stirring the two chemicals together at 80°C until a colourless, homogenous liquid is formed [4]. 20mM of anhydrous chloride salts, with purities higher than 98%, of $SnCl_2$ and $ZnCl_2$ were dissolved in the ionic liquid to provide the ions of interest. A copper film of thickness 100 nm was deposited on a glass substrate. The copper/glass substrates were cleaned in acetone in an ultrasonic bath and rinsed thoroughly with iso-propanol followed by de-ionized water. Finally, the films were deposited at constant potential mode.

Cyclic voltammetry was conducted to determine the reduction potentials of the metal salts using an Autolab PGSTAT 302N potentiostat controlled with a General Purpose

Electrochemistry Software (GPES) programme. A three-electrode system was used with a silver/silver chloride (Ag/AgCl) as the reference electrode and a platinum (Pt) foil as a counter electrode. A post-deposition thermal annealing process was performed in sulfur vapor for the sulfurization and improvement in the crystallinity of the films. The annealing process was carried out in a quartz furnace tube at 450°C for 2 hours using Argon as the carrier gas. Sulfur powder with purity of 99.998% was placed on the side of Argon gas inflow and the temperature at the position was measured to be around 150°C.

DISCUSSION

The voltammetry measurement at scanning rate $20mVs^{-1}$ showed that the peak reduction potentials for Sn and Zn in the ionic liquid are -0.55V and -1.1V respectively with Ag/AgCl as the reference electrode. A continuous film with the designed stoichiometric ratio of Cu:Sn:Zn = 2:1:1 was obtained under a constant potential at -1.15V. After the sulfurization process, the thin film exhibits a dark grey color. SEM micrograph (Fig. 1a) shows that it is a polycrystalline film with a relatively uniform surface morphology. The cross sectional SEM (Fig. 1b) shows that the structures are densely packed with an average thickness around 3 microns.

Fig. 1: SEM micrographs of CZTS showing a) top view and b) cross-sectional view.

The X-ray diffraction results shown in Fig. 2 indicated that the (101), (112), (200), (312) and (332) peaks are correlated to CZTS materials, which is in excellent agreement with the literature data (JCPDS 26-0575). Moreover, the films show the strongest peaks at (112) and (220), which characterizes a stannite structure with the preferred grain orientation along (112).

The result is different from what Katagiri *et al.* [5] previously reported using electron beam evaporation, of which the CZTS was found to be of kesterite structure. In addition our results indicate no extra phase of oxides or hydroxides in the XRD spectra.

In order to evaluate the optical properties, the transmittance of the CZTS films was measured from 500 to 990 nm. The results of the absorption measurements shown in Fig. 3 suggested that the bandgap energy of the films is around 1.49eV, which is close to the theoretical optimum value for a single-junction solar cell.

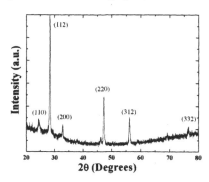

Fig. 2: XRD patterns of the CZTS thin films.

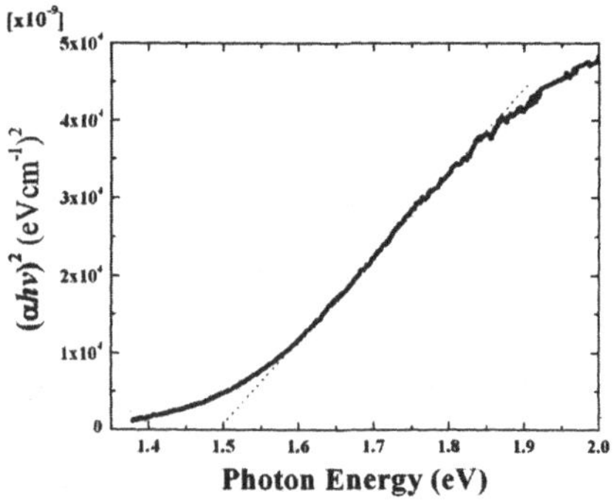

Fig. 3: Relation between squared absorption coefficient and the incident photon energy of the CZTS film.

The absorption coefficient in the visible region is in the order of $10^4 cm^{-1}$. Electrical measurements revealed that the film is a p-type semiconducting material with bulk carrier concentration of $1.7 \times 10^{19} cm^{-3}$.

CONCLUSIONS

We report a novel technique for the growth of copper zinc tin sulfide, Cu_2ZnSnS_4, thin films by electrodeposition in choline chloride based ionic liquid. The XRD results exhibit strong peaks at (101), (112), (200), (312) and (332) indicating a stannite structure. The optical absorptions results demonstrate a bandgap of 1.49 eV with an absorption coefficient of 10^{-4} cm^{-1}. Our XRD results also showed an absence of the oxide peaks in the material which is typically found in CZTS films grown in aqueous solutions. Further development of CZTS absorbers will focus on fabricating solar cell devices and investigating the conversion efficiencies of the device.

ACKNOWLEDGMENTS

This work was supported in part by grants from the Research Grants Council of Hong Kong (Project no. PolyU 5269/07E). Further support is provided by a University Research Grant of the Hong Kong Polytechnic University.

REFERENCES

1. R. Klenk and M. Ch. Lux-Stiner, "Chalcopyrite based solar cells," *Thin Film Solar Cells*, ed. J. Poortmans and V. Arkhipov (Wiley, 2006) pp.236-275.

2. I. Repins, M. A.Contreras, B. Egaas, C. DeHart, J. Scharf, C. L. Perkins, B. To and R. Noufi, *Progress in Photovoltaics: Research and Applications* **16**, 235 (2008).

3. K. N. Marsh, J. A. Boxall and R. Lichtenthaler, *Fluid Phase Equilibria* **219**, 93-98 (2004).

4. S. Z. EI. Abedin and F. Endres, *ChemPhysChem* **7**, 58-61 (2005).

5. H. Katagiri, K. Saitoh and T. Washio, *Solar Energy Materials & Solar cells* **65**, 141-148 (2001).

Mater. Res. Soc. Symp. Proc. Vol. 1123 © 2009 Materials Research Society   1123-P05-20

CdTe Films on Mo/Glass Substrates: Preparation and Properties

Vello Valdna, Maarja Grossberg, Jaan Hiie, Urve Kallavus, Valdek Mikli, Rainer Traksmaa and Mart Viljus

Department of Materials Science, Tallinn University of Technology, 5 Ehitajate Rd., EE 19086 Tallinn, Estonia

ABSTRACT

Short-bandgap group II-VI compound cadmium telluride is widely used for the infrared optics, radiation detectors, and solar cells where p-type CdTe is needed. p-type conductivity of CdTe is mainly caused by the chlorine-based A-centers, and in part, by the less stable copper-oxygen complexes. As a rule, CdTe films are recrystallized by the help of a cadmium chloride flux that saturates CdTe with chlorine. In chlorine-saturated CdTe A-centers are converted to isoelectronic complexes that cause resistivity increasement of CdTe up to 9 orders of magnitude. Excess copper and oxygen or group I elements as sodium also deteriorate the p-type conductivity of CdTe like excess chlorine. p-type conductivity of CdTe can be restored e.g. by the vacuum annealing which removes excess chlorine from the film. Unfortunately, treatment that betters p-type conductivity of the CdTe film degrades the junction of the superstrate configuration cells. In this work we investigate possibilities to prepare p-type CdTe films on the molybdenum coated glass substrates. Samples were prepared by the vacuum evaporation and dynamic recrystallization of 6N purity CdTe on the top of Mo-coated glass substrates. Then samples were recrystallized with cadmium chloride flux under tellurium vapour pressure. Results of the test studies on the structure and electronic parameters of samples are presented and discussed.

INTRODUCTION

Short-bandgap group II-VI compound cadmium telluride is widely used for the infrared optics, radiation detectors, and solar cells where p-type CdTe is needed. p-type conductivity of CdTe is mainly caused by the chlorine-based A-centers $V_{Cd}Cl_{Te}$ [1], and in part, by the less stable copper-oxygen complexes [2]. Copper co-dopant in CdTe:Cl increases the p-type resistivity value of CdTe up to some orders of magnitude and supress the efficiency of cell [3]. Copper together with oxygen increases the resistivity of CdTe:Cl up to some orders of magnitude and causes photoconductivity of films [4]. Formed at higher chlorine concentrations complex $V_{Cd}2Cl_{Te}$ is a neutral defect [5] which decreases the efficiency of cell. Attempts to use group V

elements P, As and Bi for p-type doping in CdTe did not give expexted results due to the amphotheric properties of these elements in CdTe. Reported high hole concentration of 5 x 10^{19} cm^{-3} obtained with phosphorus ion implantation [6] has been unstable. Solar cells with Bi doped CdTe exhibit a maximum efficiency of 8.0% [7], less than Cl doped ones. Codopant As rather decreases than enhances the performance of Cl doped cells [8]. Yet the best efficiency of CdTe-based cells around 16% has been obtained on chlorine doped CdTe [9] whereas the theoretically obtainable efficiencies are around 30% [10]. The CdTe-based solar cells did not exhibit any efficiency degradation even if the cells are kept under 10 suns at 100 ^{0}C [11] and are highly stable under proton flux that enables to use these cells for space applications [12].
For terrestrial power applications, ordinary soda-lime glass provides a convenient and inexpensive growth substrate [13]. For our experiments, we selected Mo-coated glass substrates [14] which are well suitable to built up substrate configuration cell structures where independent heat treatments of CdTe absorber film are possible.
In this work we investigate possibilities to prepare p-type CdTe films on the molybdenum coated glass substrate. Our goal is to use for p-type doping low chlorine concentrations comparable to the concentration of cadmium vacancies in CdTe. Results of the test studies on the structure and electronic parameters of samples are presented and discussed.

EXPERIMENTAL PROCEDURE

The CdTe films were deposited onto the Mo-coated 5x10x1 mm soda-lime glass substrates by vacuum evaporation and dynamic recrystallization (DRC) at temperatures up to 580 ^{0}C using the 6N purity degree CdTe powder. This process enabled us to form a chlorine-free CdTe film. Mechanical pressure was created by a spring, which enabled loadings up to 1000 N/cm^2. DRC processes were carried out during 15 minutes, except in the thin film experiment where the duration of 5 minutes was used. The thickness of the prepared CdTe films was between 2 and 7 microns. For the chlorine doping, 5N purity $CdCl_2$ aqueous solutions of 0.01-0.002 mole concentrations were applied on the surface of CdTe film followed by drying at 70 ^{0}C. Heat treatments of the $CdCl_2$ coated films were carried out in the Lindberg/Blue M three-zone tube furnace; isothermal recrystallization up to 30 minutes in the evacuated quartz tubes at 520 ^{0}C, followed by the treatment under tellurium vapor pressure of 0.13 kPa (1 torr) and by the vacuum annealing, with one end of the tube at 300 K. To prevent the sublimation of CdTe films during vacuum treatment, a pure CdTe monocrystal was located near the samples.
The structure of the prepared CdTe films was examined with the Jeol JSM-840A scanning electron microscope. Chemical composition and possible phase changes in CdTe films were evaluated by EDS analysis in the link Analytical AN10000 analyser. Phase changes in the DRC-processed and $CdCl_2$ treated CdTe films were investigated with Bruker AXS D5005 X-ray diffractometer.

RESULTS AND DISCUSSION

DRC process temperature of 450 ^{0}C at pressure of 400 N/cm^2 is enough to form a CdTe film, but whole film is quite porous. Higher temperature of 500 ^{0}C at pressure of 400 N/cm^2 densificates the film. Good results can be obtained at 550 ^{0}C, 400 N/cm^2 and process duration of 15 minutes (Fig. 3). Film has a dense surface without significant porosity, a columnar structure with 0.5-1 micron rod width and thickness of 7 mm. Higher temperature of 580 ^{0}C and a short

Figure 1. CdTe film on Mo-glass, formed by the dynamic recrystallization of 5 minutes at temperature T = 580 ^{0}C, ovepressure p = 400 N/cm^2. Film thickness d = 2.3 microns (sample 1).

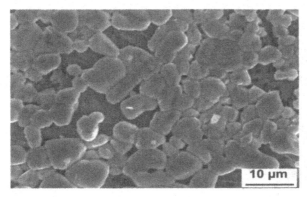

Figure 2. Sample 1 after recrystallization of 30 minutes with CdCl$_2$ flux at T = 520 ^{0}C.

process duration of 5 minutes gives a particle size around 0.2-0.3 microns, film thickness about 3 microns (Fig. 1). It seems that an optimum DRC process temperature for the soda-lime glass substrates is around 520-530 ^{0}C. All DRC samples are high-ohmic, its measured sheet resistance is around 10^8-10^9 Ohm/□.

Recrystallization with CdCl$_2$ flux significantly increases the particle size of films. A liquid phase concentration we used is too high for the thinner film and causes pinholes in the film (Fig. 2). For the thicker film used CdCl$_2$/liquid phase quantity is nearly optimum and gives a good dense film structure without pinholes (Fig. 4). The resistivity of all chlorine doped DRC films is decreased more than 6 orders of magnitude. Its sheet resistance is around 10^2 Ohm/□.

113

Figure 3. CdTe film on Mo-glass, formed by the dynamic recrystallization of 15 minutes at temperature T = 550 ^{0}C, ovepressure p = 400 N/cm^2. Film thickness d = 6.7 microns (sample 2).

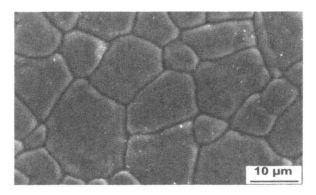

Figure 4. Sample 2 after recrystallization of 30 minutes with $CdCl_2$ flux at T = 520 ^{0}C.

This is probably a breakthrough in the CdTe-based solar cells technology. Exactly this result we have been predicted earlier [15] and hoped to receive in this work. The chlorine concentration we used in this work is nearly equal to the concentration of cadmium vacancies in CdTe film. This means that every substituted chlorine atom forms A-centre $V_{Cd}Cl_{Te}$ which is known as the most effective and stable p-type dopant in CdTe [16]. If the quantity of chlorine atoms exceeds the quantity of cadmium vacancies in CdTe, excess chlorine atoms form isoelectronic complexes $V_{Cd}2Cl_{Te}$ decreasing the concentration of A-centers.

All formed CdTe films have a face centered cubic structure with preferable orientation parallel to the (111) surface (Fig. 5). We did not detect any significant difference between the XRD spectra of DRC or chlorine doped CdTe films.

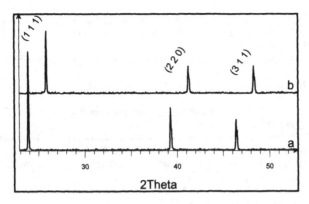

Figure 5. XRD spectra of the fabricated with a dynamic recrystallization CdTe films on the top of Mo-glass. a) Recrystallized with $CdCl_2$ flux sample. b) The same sample before $CdCl_2$ treatment.

SUMMARY

Influence of the chlorine dopant onto the structure and electronic properties of the CdTe films deposited on the top of Mo-coated glass substrates have been investigated. To avoid chlorine saturation of CdTe films were prepared first by the help of chlorine-free dynamic recrystallization process. Then, fabricated films were recrystallized and doped with chlorine using 0.01-0.002 Mole concentration $CdCl_2$ aqueous solutions. Chlorine doping at chlorine concentrations comparable to the concentration of cadmium vacancies in CdTe decreases the p-type resistivity of films up to 6 orders of magnitude. This result is probably a breakthrough in the technology of CdTe-based solar cells and supports our earlier suppositions that an efficient p-type doping of CdTe can be formed using only chlorine-based A-centers $V_{Cd}Cl_{Te}$.

ACKNOWLEDGEMENTS

This research was supported by the Bruker AXS, Inc., and Estonian Science Foundation, Grants Nos. 7241 and 7608.

REFERENCES

1. V. Valdna and J. Hiie, Proceedings of the 17th European PVSEC, Munich, Germany, 2001, pp. 1233-1235.
2. V. Valdna, Solar Energy Materials & Solar Cells **87**, 369 (2005).
3. V. Valdna, J. Hiie and A. Gavrilov, Solid State Phenomena **80-81**, 155 (2001).
4. V. Valdna, Thin Solid Films **387**, 192 (2001).
5. B.K. Meyer and W. Stadler, Journal of Crystal Growth **161**, 119 (1996).
6. H.L. Hwang, Klaus Y.J. Hsu, H.Y. Ueng, Journal of Crystal Growth **161**, 73 (1996).

7. O. Vigil-Galàn, E. Sànchez-Meza, C.M. Ruiz, J. Sastrè-Hernàndez, A. Morales-Acevedo, F. Cruz-Gandarilla, J. Aguilar-Hernàndez, E. Saucedo, G. Contreras-Puente, V. Bermùdez, Thin Solid Films **515**, 5819 (2007).
8. V. Barrioz, S.J.C. Irvine, E.W. Jones, R.L. Rowlands, D.A. Lamb, Thin Solid Films **515**, 5808 (2007).
9. M. Hädrich, N. Lorenz, H. Metzner, U. Reislöhner, S. Mack, M. Gossla, W. Witthuhn, Thin Solid Films **515**, 5804 (2007).
10. Hartley, S.J.C. Irvine, D.J. Cole-Hamilton, N. Blacker, Proceedings of the 16th European PVSEC, Glasgow, UK, 2000, pp. 816-819.
11. N. Romeo, A. Bosio, V. Canevari, A. Podestà, S. Mazzamuto and G.M. Guadalupi, Proceedings of the 19th European PVSEC, Paris, France, 2004, pp. 1718-1721.
12. Romeo, D.L. Bätzner, H. Zogg and A.N. Tiwari, Proceedings of the17th European PVSEC, Munich, Germany, 2001, pp. 1043-1046.
13. Ilvydas Matulionis, Sijin Han, Jennifer A. Drayton, Kent J. Price, and Alvin D. Compaan, Mat. Res. Soc. Symp. Proc. **668**, H8.23.1 (2001).
14. N. Romeo, A. Bosio, V. Canevari, M. Terheggen, L. Vaillant Roca, Thin Solid Films **431-432**, 364 (2003).
15. V. Valdna, Solid State Phenomena **67-68**, 309 (1999).
16. V. Valdna, Mat. Res. Soc. Symp. Proc. **607**, 241 (2000).

Mater. Res. Soc. Symp. Proc. Vol. 1123 © 2009 Materials Research Society   1123-P06-03-F07-03

Wei Liu,[1] David B. Mitzi,[1] S. Jay Chey[1], Andrew Kellock[2]

[1] IBM T. J. Watson Research Center, P. O. Box 218, Yorktown Heights, NY 10598
[2] IBM Almaden Research Center, 650 Harry Rd, San Jose, CA 95120
Email: liuwe@us.ibm.com

ABSTRACT

With tunable bandgap and demonstrated high efficiency, the chalcopyrite $CuInSe_2$ and its alloys have shown great potential as absorbers for single and multi-junction solar cells. However, the current deposition techniques mostly rely on expensive vacuum-based processing or involve complicated precursor solution preparation. These higher-cost absorber preparation processes make it difficult to commercialize this technology. In this work, $CuInSe_{2-x}S_x$ (CIS) absorbers are deposited using a simple hydrazine-based solution process. Precursor solutions were prepared by dissolving the component metal chalcogenides and chalcogen in hydrazine, forming homogeneous solutions containing adjustable concentrations of desired elements mixed on a molecular level. These precursor solutions are then spin coated on substrates followed by a heat treatment in an inert environment to produce high quality CIS thin films. Significantly, no post deposition selenization process is required using this technique. Laboratory scale devices with conventional glass/Mo/CIS/CdS/i-ZnO/ITO structure have been fabricated using CIS absorbers deposited via this process. For the baseline low-bandgap CIS system with no Ga added (to compare with our previously reported results with Ga incorporated), AM1.5G conversion efficiency of as high as ~9% has been achieved for devices with 0.45 cm² effective area.

INTRODUCTION

To make $CuInSe_2$ and its alloy a feasible photovoltaic (PV) technology, traditional vacuum-based deposition techniques face many challenges including high cost and low throughput. Current solution-based techniques attempt to address these issues, but often run into the same issues. For example, in some cases, multiple steps are required to prepare the precursor particles for solution-based deposition and undesirable oxides have to be introduced into the precursor and later taken out of the system at the cost of an extra reduction and/or selenization process [1]. With hydrazine as solvent [2-4], we can develop a much simpler process and only incorporate the necessary components of $CuIn(Se,S)_2$ into the precursor. For this method, precursor solutions were prepared by dissolving relevant metal chalcogenides and chalcogen in hydrazine, producing homogenous solutions with desired elements mixed on a molecular level. CIS thin films with good crystallinity are demonstrated by simply spin-coating the mixed precursor solutions onto Mo-coated soda-lime glass under an inert atmosphere, followed by a simple heat treatment on a hot plate. PV devices with glass/Mo/CIS/CdS/i-ZnO/ITO structure were fabricated and with efficiencies of as high as ~9% being achieved for devices with 0.45 cm² effective area. Compared with our previously published results on high efficiency solution-processed CIGS devices [5], higher currents and lower open circuit voltages were observed for these CIS devices and the overall efficiency was comparable.

EXPERIMENT

Due to the highly toxic and potentially explosive nature of hydrazine, all procedures described here should be performed with appropriate protective equipment to avoid contact with either the liquid or vapors. Unless specified, the following operations, including solution preparation, spin-coating and annealing, were all performed in a nitrogen-filled glove box with water and oxygen levels maintained below 1 ppm.

Although a single final precursor solution with constant ratios between each of the elements can be readily prepared, to conveniently adjust the Cu:In ratio within the final solution and in the final CIS film, two separate precursor solutions were prepared. Solution A containing Cu_2S was obtained by mixing Cu_2S (0.9549 g, 6 mmol), S (0.3848 g, 12 mmol) and anhydrous hydrazine (12 mL), leading to a clear yellow solution after several days of stirring. Solution B (In_2Se_3) was prepared by mixing In_2Se_3 (1.399 g, 3 mmol), Se (0.2369 g, 3 mmol) and anhydrous hydrazine (12 mL), yielding a viscous colorless solution.

The final spin-coating solution was prepared by mixing Solutions A and B in an appropriate ratio to obtain the targeted stoichiometry of the CIS film. Optional extra Se can be added into the solution to compensate the escaped Se during annealing. Extra hydrazine can also be added to obtain a targeted precursor solution concentration. For example, targeting a $Cu_{0.8}In(Se,S)_2$ stoichiometry with an approximately 0.29 mole CIS per liter of hydrazine concentration, 0.8 mL solution A (0.8 mmol Cu) was mixed with 2 mL solution B (1.0 mmol In) and 0.7 mL hydrazine. Optionally, 0.5 mmol extra Se per mmol of CIS can be added and allowed to dissolve.

The spin-coating solution was filtered through a 0.2 μm PTFE filter and flooded onto the substrate, which was then spun at 800 rpm for 90 seconds. For device purposes, thick CIS (on the order of 1 μm) are required to sufficiently absorb most of the incident photons. To achieve this thickness, multiple deposition cycles are used. In between successive deposition cycles, the substrates were subjected to an intermediate anneal on a hot plate at 290 °C for 5 min, which drives away hydrazine solvent and forms poorly soluble CIS, preventing re-dissolution during the next cycle. Under the conditions described in this paper, 6-10 cycles of deposition may be needed to produce > 1 μm thick CIS. After all depositions were finished, a final high temperature annealing was implemented to drive away any residue hydrazine and extra chalcogen species and to allow grain growth within the CIS thin films.

All devices fabricated adopted a substrate structure. CIS thin films were deposited on 2.54×2.54 cm^2 soda-lime glass pieces (0.1 cm thickness) coated with approximately 750 nm of sputtered Mo. CIS films for device purpose were spin-coated 8 times using the process described above to achieve a thickness of approximately 1.6 μm (as verified using SEM analysis of a film cross section). Final annealing was done at 500 °C for 20 minutes. The n-type junction partner CdS (approximately 50nm thick) was deposited using a chemical bath deposition similar to what is described elsewhere [6]. The CdS side of the substrate was then sputter coated with an approximately 70-nm-thick layer of intrinsic ZnO followed by 180 nm of transparent conductive indium tin oxide. The optical transmission of the ITO layer was maintained above 80% in the visible spectral range while the sheet resistance was as low as ~40 Ω/ψ. Finally, a Ni (50 nm)/Al (2 μm) dual layer top contact grid was evaporated to finish the device.

For each set of devices prepared, sister CIS samples were deposited. For X-ray diffraction, Mo-coated glass was used as substrate and for RBS compositional analysis, thermally oxidized Si was used instead. In the case that the CIS film did not attach well on Si after annealing, samples were deposited on Mo coated glass substrates for PIXE analysis. Powder XRD data were collected using a Siemens D5000 x-ray diffractometer with Cu Kα radiation. Scanning electron microscopy (SEM) pictures were taken using the cross-section of freshly cleaved CIS samples, which were coated with a thin Pd–Au film to prevent electron charging. Rutherford backscattering spectrometry (RBS) was performed using an NEC 3UH Pelletron, with a beam current of 20 nA @ 2.3 MeV. PIXE was performed using the same machine with 1.1 MeV protons.

The finished devices were electrically tested using a Newport Oriel Solar I-V system which consists of a solar simulator equipped with an Air Mass 1.5G filter. The light source was calibrated to 1 sun using an Oriel reference cell calibrated and certified by NREL. Field measurement on a cloudless day was also performed to confirm the measured efficiency of these devices.

DISCUSSION

Complete CIS solar cells are fabricated and characterized for I-V and quantum efficiency properties. Films are also characterized by XRD, RBS and cross-section SEM.

X-ray diffraction

Figure 1 shows the X-ray diffraction data of a CIS thin film deposited on a Mo-coated glass substrate. The sample was subjected to a final anneal at 500 °C for 20 minutes in a nitrogen atmosphere. Good crystallinity is noted in the X-ray data and the grains are preferentially (112) oriented. The broad peaks indicated by asterisks are a Mo/Se phase formed at the interface between Mo substrate and $CuInSe_2$ film [7].

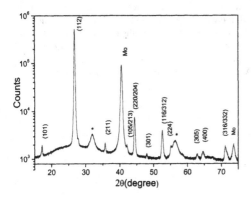

Figure 1. Powder X-ray diffraction data of a $CuInSe_2$ film deposited on a Mo-coated glass substrate, showing (112) preferred orientation (* indicates interfacial $MoSe_2$ phase).

Device characterization

Figure 2 presents the light (AM 1.5G spectrum) and dark current vs. voltage characteristics of two different devices (E2 and A) with CIS absorbers processed using the process described earlier. Efficiencies of up to ~9% were achieved on these devices.

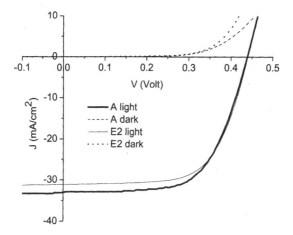

Figure 2. I-V characterization of two CIS devices (E2: Efficiency = 8.86%, V_{oc} = 0.440 V, J_{sc} = 31.2 mA/cm^2, FF = 64.6%; A: Efficiency = 9.13%, V_{oc} = 0.439 V, J_{sc} = 33.0 mA/cm2, FF = 63.0%).

External quantum efficiency data were taken on both devices and the results are shown in Figure 3. Due to the limitation of the tool, the wavelength data beyond 1100 nm was not available. The devices have good current collection for wavelengths below 750nm, while at longer wavelengths the quantum efficiency starts to drop. The better collection efficiency of sample A over sample E2 within the wavelength region between 400 nm and 500 nm is mainly due to the thinner CdS layer of sample A, causing less absorption in the buffer layer.

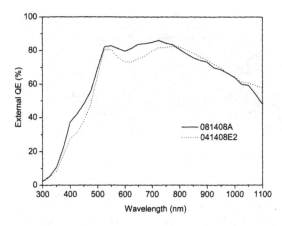

Figure 3. QE characterization of the two CIS devices shown in Figure 2.

Final film compositions of the two CIS samples measured by RBS or PIXE are shown in Table I. In the case of PIXE analysis, S content was not distinguishable due to peak overlapping. Both films bear a Cu poor composition.

Table I. CIS film compositions measured by RBS/PIXE

Sample	[Cu] at.%	[In] at.%	[Se] at.%	[S] at.%
081408 A	21.1±0.5	26.4±0.5	47.4±0.5	5.1±0.5
032708 E2	20.4±0.5	22.9±0.5	56.7 ±0.5	N/A

Grain structure characterization

In analogy to vacuum-grown CIS, it is generally believed that large grain size leads to less carrier recombination and therefore higher current collection. Scanning electron microscopy was employed to evaluate grain structure and, together with RBS composition analysis, a correlation between composition and grain structure was established. As shown in Figure 4A, large grain size, on the order of microns, can be readily obtained using this hydrazine precursor approach. It is found that this large grain structure correlates to a Cu-rich composition (Cu:In ratio>1), which unfortunately leads to poor performance devices due to adverse shunting. Figure 4B shows the grain structure of a Cu-poor CIS device. The grain size is smaller compared to the Cu-rich ones, but still maintains a reasonable size of a few hundred nanometers. So far, high-efficiency devices all bear a Cu-poor composition. Note that, recently, hydrazine-processed CIS devices, deposited on silicon (rather than glass) and with an efficiency of 3.5% (0.13 cm^2 device area), have also been reported by Hou *et al*. [8].

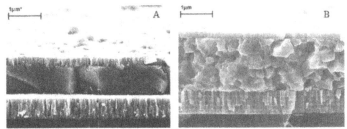

Figure 4. SEM cross-section image of CIS devices with different Cu to In ratio. Sample A: Cu:In : ~1.1; Sample B: Cu:In : ~0.9 (The top layer with small column shaped grains is CdS/ZnO/ITO).

CONCLUSIONS

CuInSe$_2$ thin films were successfully deposited using a new solution-based technique in which absorber layer deposition involves precursors dissolved in hydrazine. Good crystallinity and grain size were obtained using this hydrazine-based approach. High efficiency, approaching values for devices based on more cost-intensive vacuum-grown CIS, was achieved on devices fabricated using these solution processed absorbers. Comparing to CIGS devices prepared using the hydrazine approach [5,6], CIS devices have higher short circuit current but lower open circuit voltage due to the bandgap difference. However the overall efficiency is comparable. The ability to fabricate high-efficiency CIS solar cells further demonstrates the flexibility of the hydrazine-based approach for depositing metal chalcopyrite-based absorber layers and provides further evidence that this approach may offer a low-cost, high-efficiency route to thin-film PV device fabrication.

ACKNOWLEDGMENTS

The authors thank R. Ferlita for technical support with the preparation of the Ni/Al grid, and R. Noufi for providing the shadow mask design used for this study.

REFERENCES

1. V. K. Kapur, A. Bansal, P. Le, and O. I. Asensio, Thin Solid Films, **53**, 431–432 (2003).
2. D. B. Mitzi, L. L. Kosbar, C. E. Murray, M. Copel, A. Afzali, Nature, **428**, 299 (2004).
3. D. B. Mitzi, M. Copel, C. E. Murray, Adv. Mater., **18**, 2448 (2006).
4. D. J. Milliron, D. B. Mitzi, M. Copel, C. E. Murray, Chem. Mater., **18**, 587 (2006).
5. D. B. Mitzi, M. Yuan, W. Liu, A. Kellock, S. J. Chey, V. Deline, A. G. Schrott, Adv. Mater., **20**, 3657 (2008).
6. M. A. Contreras, M. J. Romero, B. To, F. Hasoon, R. Noufi, S. Ward, K. Ramanathan, Thin Solid Films, **204**, 403(2002).
7. D. B. Mitzi, M. Yuan, W. Liu, A. Kellock, S. J. Chey, L. Gignac, A. G. Schrott, Thin Solid Films, in press, doi:10.1016/j.tsf.2008.10.079 (2008).
8. W. W. Hou, S. Li, C. Tung, R. B. Kaner, Y. Yang, in Solar Energy: New Materials and Nanostructured Devices for High Efficiency, (Optical Society of America, 2008), paper SWC1.

Characterization

Mater. Res. Soc. Symp. Proc. Vol. 1123 © 2009 Materials Research Society   1123-P02-05

Computer Simulation of Edge Effects in a Small-Area Mesa N-P Junction Diode

Jesse Appel[1], Bhushan Sopori[1] and N.M. Ravindra[2]
[1]National Renewable Energy Laboratory, Golden, CO 80401, USA
[2]New Jersey Institute of Technology, Newark, NJ 07102, USA

ABSTRACT

The influence of edges on the performance of small-area solar cells is determined using a modified commercial, finite-element software package. The n^+/p mesa device is modeled as having a sub-oxide layer on the edges, which acquires positive charges that result in development of an electric field within the device. Our computer simulations include generation/ recombination at the diode edges as well as the influence of light on the recombination characteristics of the edges. We present a description of our model, dark and illuminated characteristics of devices with various surface charge concentrations, and the dynamics of carrier generation/recombination. The influence of edge geometry on diode performance is determined.

INTRODUCTION

Use of mesa diode arrays, fabricated by chemical etching, is a valuable technique for detailed characterization of photovoltaic properties of single-crystalline and multicrystalline silicon wafers. The chemical etching process creates a sub-oxide, which is rich in hydrogen and passivates the edges. The diode array consists of 100-mil-diameter devices that are etched 3 to 4 microns deep and have a blanket back Al contact common to all devices. Each device has a front Al contact, 10 mils in diameter, for ease in automatic probing. The diodes are electrically isolated from one another and can be probed to measure the current density versus voltage curves under dark and illuminated conditions [1,2]. We find that the degree of passivation depends strongly on the process conditions. To understand the details of the edge passivation, we modeled an n-p junction diode using a commercial, finite-element software package. These simulations have led to a determination of the self-consistent solution to the continuity equations for electrons and holes using the steady-state drift-diffusion model for carrier dynamics coupled with electric potential determined from Poisson's equation [3]. The purpose of these simulations is to determine the influence of edge conditions on the overall performance of mesa diodes under dark and illuminated conditions. In particular, we examine the effect of edge shape on the I-V characteristics of the diode.

The underlying mechanisms of bulk and surface recombination have been well established for crystalline silicon semiconductor devices [4-7]. We have applied them to our mesa device using the COMSOL software. Our simulations show that the space-charge region becomes extended along the vertical edge of the mesa diode due to the fixed positive surface charge. At the intersection of the vertical edge and step, a strong electric field is produced because it has a small convex radius of curvature [8]. Depending on the sharpness of this intersection, the entire device can become significantly shunted. Simulations have been performed with a sharp corner and a smooth curve at the intersection of the vertical edge and the step. The use of a smooth curved transition results in significantly lower dark current density versys voltage and a greater open-circuit voltage and fill factor under illumination. Yet, even with a curved transition, the space charge region can extend approximately 100 microns into a 199.5-micron-thick mesa diode and have a bulk recombination rate that is two orders of magnitude greater than the rest of the device at low forward biases.

SIMULATION RESULTS

The diode configuration used for modeling is illustrated in Figure 1. Each diode is 100 mils in diameter, etched 3.5 μm deep, and has a concentric front metallization of 10 mils. The surface of the diode is coated with an antireflection layer having an interface charge of $1x10^{11}$ cm^{-2}. In the current simulation, we assume that the edge of the diode has a thin oxide, and the charge density of the oxide is varied from $1x10^{11}$ cm^{-2} to $5x10^{11}$ cm^{-2}. The junction depth is assumed to be 0.5 μm.

The mesa diode was simulated under dark and illuminated conditions. Figure 1 shows a cross section of the device, including the lifetime and surface recombination velocity parameters used in all of the simulations. Detailed simulations have shown that charge loss can occurs via a number of mechanisms, which include: carrier flow from the diode to the edge, recombination by the field produced at the edge, which propagates radially into the diode and through the thickness of the diode, and recombination at the etched surface.

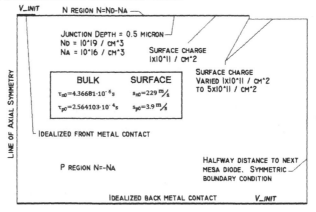

Figure 1. Cross section of modeled mesa diode with bulk and surface conditions

Dark Condition

The dark current density versus voltage plots of the simulations are shown in Figure 2. The current density increases with increasing surface charge. The initial increase in dark current density is due predominantly to recombination near the mesa's edge. The greatest increase occurs at low voltages and is due to resistive shunting, which is related to the distortion of the electric field at the edge of the mesa. One of the most important results of the simulations performed on the mesa diode is the calculation of a high point in the electric field at the corner of the vertical edge and the step. The resultant current density versus voltage curves for the simulations in Figure 2 were performed with a corner that was a right angle with a sharp point. To determine if this feature of the mesa diode was responsible for the increase in dark current density, similar simulations were run with a curved transition rather than a sharp corner. A small, curved piece of p-type semiconductor was added at the corner of the vertical edge of the mesa and the step. It was defined as a second-order Bezier curve with a height and length of 3.5 μm. This height was chosen because the junction is 3.5 μm above the step. The bulk

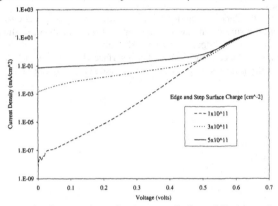

Figure 2. Dark current density vs. voltage for the mesa diode at different vertical edge and step surface charges

recombination rate and electric field for simulations with the vertical edge and step surface charges of 5×10^{11} cm^{-2} are shown for sharp and curved corners in Figure 3. The addition of a curved piece at the mesa's edge clearly reduces the bulk recombination rate in this region. However, the curved piece has a smaller radius of curvature than the flat continuous sections away from it. Therefore, even with an additional curved piece, the bulk recombination rate

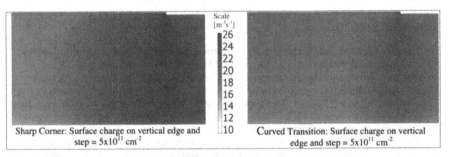

Figure 3. Comparison of bulk recombination rate (logarithmic) and electric field at 0.1 volt forward bias for a mesa diode with sharp and curved corners with a surface charge of 5×10^{11} cm^{-2} under dark conditions.

127

and electric field in this region are greater than the rest of the device except for the junction, where they are almost equal. Consequently, the results of the dark simulations of the mesa diode indicate that the distortion of the electric field at the corner of the junction and the vertical edge, and the step and the vertical edge in conjunction with the inversion layer on the p-side adjacent to the junction, results in an extension of the space-charge region deep into the bulk of the device. This leads to an increase in the recombination rate and the dark current, particularly at low forward biases.

A comparison of the dark-current density versus voltage for sharp and curved corners is shown in Figure 4 for a surface charge of 5×10^{11} cm^{-2}. By increasing the radius of curvature of corner at the step and vertical edge, the resistive shunting caused by excessive bulk recombination can be reduced significantly. For a surface charge of 5×10^{11} cm^{-2}, the reduction of dark current density is almost four orders of magnitude at low forward biases.

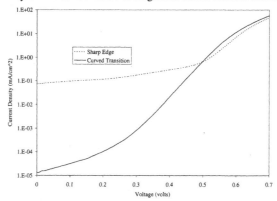

Figure 4. Current density vs. voltage plots of the comparison of sharp and curved transitions between the step and the vertical lines with a surface charge of 5×10^{11} cm^{-2} under dark conditions.

Illuminated Condition

Simulations of the mesa diode were also performed under illuminated conditions, in which only the vertical-edge surface charge was changed from 1×10^{11} cm^{-2} to 5×10^{11} cm^{-2}. The resulting illuminated current density versus voltage curves are shown in Figure 5. The open-circuit voltage and fill factor become reduced as the vertical-edge surface charge is increased from 1×10^{11} cm^{-2} to 5×10^{11} cm^{-2}. The loss in V_{oc} and FF are significant, and the mechanism that causes them as the charge increases is a result of increased recombination due to the increased electric field caused by the sharp corner at the edge of the mesa.

To determine if the sharp corner at the edge of mesa was the cause of the reduction in open-circuit voltage and fill factor, simulations under illumination were performed with a curved piece at the corner of the vertical edge and the step. It had exactly the same dimensions as the one used for the dark simulations. Also, the simulations using a sharp corner were repeated with the surface charge of the vertical edge and the step set to the same value.

128

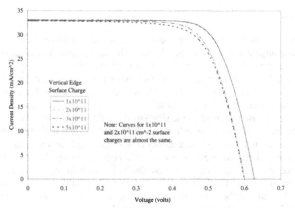

Figure 5. Illuminated current density vs. voltage for the mesa diode at different vertical edge charges.

The illuminated current density versus voltage plots for simulations performed with a sharp corner and a curved transition with a surface charge of 5×10^{11} cm^{-2} are shown in Figure 6. The simulation with a curved transition has a greater open-circuit voltage and fill factor. This occurs because there is a lower recombination rate in the mesa diodes modeled with a curved transition. It should be noted that the short-circuit current (J_{sc}) values for the simulations shown in Figure 6 are significantly greater than the J_{sc} values shown in Figure 5; this is because the surface charge was increased on the step adjacent to the mesa's edge. This results in an inverted surface, which significantly reduces the recombination rate in this region, and consequently leads to an improved J_{sc} value.

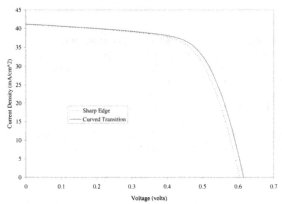

Figure 6. Current density vs. voltage plots of the comparison of sharp and curved transitions between the step and the vertical lines with a surface charge of 5×10^{11} cm^{-2} under illuminated conditions.

CONCLUSIONS

The geometric features of the edge of the mesa diode can significantly distort the electric field throughout a large portion of the bulk device. The most important feature occurs at the intersection of the vertical edge and step because an electric field becomes large near areas having a convex radius of curvature and can reach extreme values at sharp points [8]. In certain cases, the distortion can expand the space charge region over hundreds of microns in all directions emanating from the edge. It was shown that changing the interface from a sharp perpendicular step to a Bezier curve in the simulations can greatly reduce the dark-current density and increase the open-circuit voltage and fill factor. This is important because the chemical etch used to delineate the mesa diodes is isotropic and leaves behind a curved transition. Even with a curved interface at the vertical edge and step, the edge effect is a significant source of recombination in the mesa diode, in which the recombination rate can be more than two orders of magnitude greater than the rest of the bulk and this region can extend approximately 100 μm into the device from the curved interface. Therefore, the edge effect in a mesa diode is a result of the space-charge region extending along the vertical edge because of the inverted region on the p-side of the junction, then expanding deep into the bulk device because of the strong electric field at the intersection of the vertical edge and the step. This results in a significant increase in recombination rate throughout the device, an increase in dark-current density, and a reduction in open-circuit voltage and fill factor under illumination.

ACKNOWLEDGEMENTS

We would like to thank Peter Rupnowski for his help with calculations and many valuable discussions.

REFERENCES

1. B.L. Sopori, *Applied Physics Letters* **52** (20), 1718-1720 (1988).
2. B.L. Sopori, *Journal of Applied Physics* **64** (10), 5264-5266 (1988).
3. H.K. Gummel, *IEEE Transactions on Electron Devices* **11**, 455-465 (1964).
4. R.N. Hall, *Physical Review* **87**, 387 (1952).
5. W. Shockley and W.T. Read, Jr., *Physical Review* **87** (5), 835-842 (1952).
6. A.S. Grove, B.E. Deal, E.H. Snow and C.T. Sah, *Solid-State Electronics* **8**, 145-163 (1965).
7. D.J. Fitzgerald and A.S. Grove, *Surface Science* **9**, 347-369 (1968).
8. Raymond A. Serway, *Physics for Scientists & Engineers*, 3rd ed. (Saunders College Publishing, Philadelphia, Vol. II, p. 692 (1992).

Quantitative Measurement of Boron and Phosphorus in Solar Grade Silicon Feedstocks by High Resolution Fast-Flow Glow-Discharge Mass Spectrometry

Karol Putyera[1], Kenghsien Su[1], Changhsiu Liu[1], R. S. Hockett[2] and Larry Wang[2]

[1]Evans Analytical Group - New York, 6707 Brooklawn Parkway, Syracuse, NY 13211, U.S.A
[2]Evans Analytical Group - California, 810 Kifer Road, Sunnyvale, CA 94086, U.S.A.

ABSTRACT

The calibration factors are examined for determination of boron (B) and phosphorus (P) in solar grade silicon samples in the new generation of high resolution fast-flow glow-discharge mass spectrometers (FF-GDMS). It is shown that using the generalized calibration factors from the Standard RSF table, the relative errors observed in the determination of these analytes is not acceptably small for photovoltaic applications. The certified B and P values in NIST SRM 57a Si metal sample do not have the confidence intervals, which would be adequate for refining these calibration factors to the acceptable levels. Thus, well-characterized single crystalline Si wafers with known boron and phosphorus contents traceable to NIST reference materials (SRM 2133 and SRM 2137) were used for accurate calibrations and for establishing good analytical procedures for measurements of wide variety of Si sample forms. The obtained results for these important analytes using the new FF-GDMS procedure are compared to other characterization techniques commonly used in this industry.

INTRODUCTION

The most common impurities of interest in photovoltaic Si feedstocks and wafers are the dopants (mainly B and P), the atmospherics (mainly O for p-type Si, and C), and the transition metals (mainly Fe). A wide range of analytical techniques are used for detecting impurities in PV Si, some more appropriate than others depending on the purity level and on the solar cell design, but each with its strengths and weaknesses [1].

The lack of adequate Standard Reference Materials and different use of unverified secondary standards, may lead to defective Quality Control and Quality Assurance of analytical methods. This, in addition to difficulties with calibrations, may lead to systematic errors in determinations of important analytes. Sample preparations, sample handling and/or pre-treatment prior to analysis using the most common analytical techniques are also identified as the main sources in measurement errors.

There is a need for analytical technique(s), which can deliver accurate data to allow close process and production controls. Two analytical techniques which require little to no sample preparation that may introduce or remove impurities of interest and which have excellent detection limits in Si are Secondary Ion Mass Spectrometry (SIMS) and High Resolution Glow Discharge Mass Spectrometry (GDMS) [2, 3, 4]. GDMS now has an international standard test method for solar grade silicon under SEMI [5].

The new generation of GDMS instruments (Thermo Fisher Scientific, Model Element GD) is a combination of a fast-flow glow-discharge ionization source (Figure 1) with a high resolution mass analyzer. The FF-GD is a high power ionization source resulting in high atomization/sputter rates. Consequently, the atomization rates are on the order of microns per

131

minute for metals in the FF-GD as compared to 0.01-0.1 microns per minute in the previous generation of GDMS instruments.

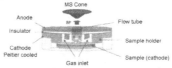

Figure 1: Schematics of the fast-flow glow discharge ion source.
This is the first instrument of its kind designed for fast determinations of trace to ultra-trace impurities in variety of solids including high purity Si samples for the PV industry [6]. In the present work, the setup of this new FF-GDMS instrument for trace to ultra-trace level determinations of B and P in various Si samples is presented.

THEORY

FF-GDMS is among the most sensitive analytical techniques today for direct elemental analysis of solids. One of the great advantages of this technique is that the measurements are basically always "quantitative" due to the fairly uniform sensitivities when determining most of the elements. Generally, the nominal results are within the range of approximately 2x the "true" values with the "built-in" calibration parameters. This distinguished feature of the GDMS measurements makes the technique extremely powerful especially for measurements of unknown compositions or for determinations of elements for which there are no certified reference values available, like most of the elements in Si samples. This is also a very powerful attribute for depth profiling of layered structures composed of variety of matrixes or for distribution studies of elements in micro-volumes. The fundamental principle of the "built-in" calibration parameters is related to the core of the GDMS quantification principles in general [7]. This is based on the Ion Beam Ratio (IBR) methodology. The IBR is the quantified ratio of the intensity of the analyte element signal relative to the matrix element's) signal intensities. In the case of pure Silicon measurements for example, the Si ion intensity is directly measured and the trace elements signals are rationed to the Si ion beam all determined under the same atomization/ionization conditions. In a "true" calibration setup the certified element concentrations are plotted against the measured IBRs of each element. The slope of this "calibration graph" gives the "true" sensitivity or the relative sensitivity factor (RSF) for that element. According to the above definition, the RSF for element x in Silicon matrix can be defined as:

$$C_x/C_{Si} = RSF_{x/Si} * I_x/I_{Si} ,$$

where I and C are the ion current and the concentration in mass units, respectively, and x and Si represent the element x and the internal standard Si, respectively. There are several other analytical techniques, which are based on this or similar calibration methodology. However, the element sensitive RSF values in GDMS are uniform and they are close to unity for most of the elements. Furthermore, the RSF values are varying in a relatively narrow range from element to element since they are not significantly influenced by the nature of the sample matrix. Consequently, acceptable RSF sets can be developed without having multiple or strictly matrix matched reference materials. This distinguishing feature of the GDMS technique led to development of a generalized set of RSF values, corresponding to a universal calibration curve, which has proven to be explicitly useful for measurements of unknown compositions or for materials without using certified reference values for calibrations. This generalized RSF set is frequently used as the starting calibration, from which the "true" calibration coefficients are refined, if needed, using appropriate reference materials.

RESULTS AND DISCUSSION

The B and P results for NIST SRM 57a Si metal in Table I have been measured as ion beam ratios and converted to bulk concentrations by applying the respective calibration factors from the Standard RSF table "built-in" of the GDMS instruments. This approach is common practice for GDMS analyses in general and expected to provide nominal results within 30% of the true values – semi-quantitative results.

Table I: Boron and Phosphorus GDMS results obtained on NIST SRM 57a Si metal sample.

Analyte	Certified value wt%	Estimated Uncertainty wt%	Measured Value mg/kg	RSD% (n=12)	Ratio of Measured/Certified
B	0.001	---.	9.1	24	0.91
P	0.003	0.001	24.5	30	0.82

Here we should note that the certificate of the SRM 57a declares that the estimated uncertainty should be understood that it represents an evaluation of the combined effects of method imprecision, possible systematic errors among methods and material variability for samples 0.5g or more. There is no estimated uncertainty reported for the Boron results in the certificate of this SRM. Thus, we have chosen B(HI) = 0.0014 wt% (14 mg/kg or 14 ppm wt) for the upper concentration control level and B(LO) = 0.0005 wt% (5 mg/kg or 5 ppm wt) for the lower concentration control level, respectively, i.e. the concentration values from rules of rounding.

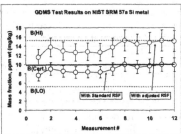

Figure 2: Comparison of twelve independent GDMS tests results for B measured on NIST SRM 57a with B RSF from the Standard RSF Table and with adjusted B RSF after re-calibrating the instrument using well-characterized traceable reference materials.

The averaged B and P results from 12 independent analyses are presented in Table I. These results indicate that the obtained concentration values are indeed fairly accurate. The relative standard deviations of results for these two important analytes are also within the expected uncertainties or control levels. Based on these measurements one could conclude, that the RSF values for B, P and Si respectively, in the instrument's Standard RSF table is "acceptable" for accurate determinations of Boron and Phosphorus in any Silicon samples. However, if the same RSF table is applied for verifications of Boron concentrations on well-characterized Si samples with known Boron values, which are traceable to NIST standard reference material SRM 2137, the GDMS results of B indicate that they are systematically biased lower as compared to the "known" or "true" values. Such experimental GDMS results are compared in Table II for four different Si reference samples (EAG Reference Materials). These reference Si samples are characterized with Boron concentrations in approximately 3 orders in the concentration scale The GDMS results are clearly biased low by about 40% to 50% on all four samples..

Table II: Comparison of GDMS results with SIMS reference values for well-characterized Si reference samples – all results on ^{11}Boron isotope.

Sample/Analyte	SIMS Result, mg/kg	RSD%	GDMS Result, mg/kg	RSD%	Ratio of SIMS/GDMS
#1/^{11}B	0.08	2	0.06	40	1.3
#2/^{11}B	0.22	3	0.16	10	1.4
#3/^{11}B	120	2	80	20	1.5
#4/^{11}B	62	1	43	20	1.4

Notes: The SIMS results are traceable to NIST Si reference material (SRM 2137 - B).

Phosphorus determinations were conducted and compared with SIMS results the corresponding way to Boron by using certified Si wafer (SRM 2133 - P) and several up-graded metallurgical grade Silicon samples. Based on these comparative tests results, the correct RSF value for Phosphorus in Si was developed. The GDMS measurements for the two most important dopants, B and P, were now considered to be traceable to NIST Standard Reference Materials.

The B and P RSF values were adjusted in order to make the GDMS results accurate, i.e. corresponding to the reference values and the SIMS results. The impact of such adjustment to the GDMS test results plot for SRM 57a for B determinations is illustrated in Figure 2. The B values originally measured with the B RSF from the Standard RSF table all are around the central horizontal line representing the "certified" B concentration in this Si SRM. However, even if the B RSF is adjusted, increased by 50% in order to remove the bias between the SIMS results and the GDMS results, the Boron results with the adjusted B RSF will still remain within the B(HI) and B(LO) control levels as shown in Figure 2. This is a good example for illustrating the specifics with GDMS calibrations for determinations of trace element concentrations in general. Generally, the estimated uncertainties of trace concentrations in Certified Reference Materials are not better then in SRM 57a. This degree of uncertainties to the 10 mg/kg (ppm wt) certified concentrations indicate, for instance, 7 mg/kg in the low point and 13 mg/kg in the high point of the control levels, which translates to a factor of approximately two between the low and high border line values. Thus, for developing/verifying the "true" sensitivity factors for trace level analyses there is a need for reference materials having not just known "true" value(s) but also stringent confidence intervals.

B and P doped solar cell wafers (200+ micron thick wafers) were used to tests the accuracy and reproducibility of the adjusted RSF table for the determination of the two most important dopant concentrations in Si. Table III shows comparison of results on six different solar wafers. Dopant concentrations on all six wafers were compared to SIMS results and on four of the wafers to four-point probe resistivity measurements converted into dopant concentrations.

Table III: Comparison of various results on six different solar wafer Si samples

Sample	Remarks	Resistivity	Cell Efficiency	SIMS B, mg/kg	SIMS P, mg/kg	GDMS B, mg/kg	GDMS RSD%	GDMS P, mg/kg	GDMS RSD%
# 1	(1)	1.5E-2		0.09	0.01	0.10	7	0.03	19
# 2	(2)	0.5E-1		0.74	1.69	0.65	10	N.S.	N.S
# 3		0.5E-1	15.0%	0.44	0.57	0.40	7	0.78	2
# 4		1.5E-2	15.4%	0.08	0.003	0.09	11	0.01	25
# 5	Good		15.4%	0.07	0.01	0.10	2.5	0.04	30
# 6	Bad		6.0%	0.11	0.02	0.78	N.S	N.S	N.S

Notes: (1) low leakage current (2) very high leakage current; N.S. - GDMS readings are not stable or not resulting in equilibrated values.

The SIMS and GDMS measurements now agree on good wafers, but they disagree on results for the "bad" ones (samples # 2 and # 6 in particular). This may caused by differences in the sampling volumes between these two techniques and/or differences in microscopic distributions of these elements in the "bad" wafers. Also, the precision or relative standard deviations of GDMS measurements for B and P were observed in the desired single digit RSD%'s only for concentrations above 100 µg/kg (or ppb wt) concentration levels.

The FF-GDMS is capable of directly determining impurities in photovoltaic Si solar cells with single digit µg/kg (ppb wt) detection limits. However, besides the sensitivity of such determinations, the dispersion of a set of data points about its central axis is also very important for PV applications, especially if the contents are in the ultra-trace concentration levels. Thus, the precision for such B and P determinations was investigated using single crystalline Si wafers from the NIST SRM 2551 series. The NIST SRM2551 contains four Si samples characterized with different oxygen contents. The aim of these tests was to investigate the precision of B and P determinations for concentrations close to the detection limits of the method routinely used for multi element determinations in Si wafers. The obtained results are summarized in Table IV.

Table IV: Precision of ultra-trace B and P determinations in various Si samples

Sample ID	Oxygen Content, mg/kg	B measured, µg/kg	RSD% (n=9)	P measured, µg/kg	RSD% (n=9)
FZ	< 0.06	5	40	15	9
CZ-L	8.5	3	50	20	16
CZ-M	(8.5-17)	4	40	6	22
CZ-H	17	6	37	14	22

Notes: FZ – Float Zone specimen; CZ – Czochralski specimens; L – low oxygen content; M – medium oxygen content; H – high oxygen content.

These results confirmed that the B and P measurements by GDMS are less precise, as compared to SIMS measurements for routine B and P determinations in the ultra-trace concentration levels. It should be noted though, that for this particular evaluations the instrumental settings were not specifically set for determinations simply of these two analytes with long integration times, but with settings that are normally used for routine multi-element analysis of Si samples in general. Consequently, the pre-sputter time was set to 10 min and the analysis time was also around 10 minutes in these measurements. Figure 3 illustrates the typical sputter crater after 20 min analysis. The sampling mass in these types of Si analyses is around 5 mg per test.

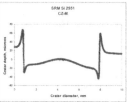

Figure 3: Typical plasma sputtered crater on SRM Si 2551 CZ-M after 20 min analysis.

CONCLUSIONS

1. FF-GDMS can provide in "semi-quantitative mode" very useful and fairly corresponding information on trace elements in Si samples.
2. Developing accurate sensitivity factors for GDMS determinations is essential in order to obtain results for QC/QA, pr process control or for electrical characterizations of solar grade Si samples for the PV industry.
3. The sampling depth in the GDMS measurements is in the order of tens of microns, thus the material has to be homogeneous on this scale for the results to be representative.
4. With a routine multi-element collection mode, GDMS measurements can lead to precise results, with single digit RSD%, if the dopants' concentrations are above the 100 μg/kg (ppb wt) levels.

REFERENCES

1. R. S. Hockett in Proceeding of the 18th Workshop on Crystalline Silicon Solar Cells & Modules: Materials and Processes, edited by B. L. Sopori, published by NREL 2008, pp. 48-59.
2. L. Wang and R.S. Hockett in Proceedings of 23rd European Photovoltaic Solar Energy Conference 2008, pp. 1209-1212
3. C. Michellon, K. Putyera, M. Kasik and R.S. Hockett in in Proceedings of 23rd European Photovoltaic Solar Energy Conference 2008, pp. 2328-2331.
4. C. Venzago, H.v. Campe and W. Warza, in Proceedings of 11th European Photovoltaic Solar Energy Conference 1992, pp. 484-486.
5. SEMI PV1-0309 - Test Method For Measuring Trace Elements in Photovoltaic-Grade Silicon by High-Mass Resolution Glow Discharge Mass Spectrometry
6. J. Hinrichs, M. Hamester and L. Rottman, Thermo Fisher Scientific, www.thermo.com, Application Note: 30164.
7. M. Di Sabatino, A.L. Dons, J. Hinrichs, O. Lohne and L. Arnberg, in Proceedings of the 22nd European Photovoltaic Solar Energy Conference 2007, pp 271-276.

Mater. Res. Soc. Symp. Proc. Vol. 1123 © 2009 Materials Research Society 1123-P07-10

Bhushan Sopori[1], Przemyslaw Rupnowski[1],
Jesse Appel[1], Debraj Guhabiswas[1], LaTecia
Anderson-Jackson[2]
[1]National Renewable Energy Laboratory, Golden, CO 80401
[2]North Carolina Agricultural and Technical State University, Greensboro, NC, 27411

ABSTRACT

We report on our observations of light-activated passivation (LIP) of Si surfaces by iodine-ethanol (I-E) solution. Based on our experimental results, the mechanism of passivation appears to be related to dissociation of iodine by the photo-carriers injected from the Si wafer into the I-E solution. The ionized iodine (I^-) then participates in the formation of a Si-ethoxylate bond that passivates the Si surface. Experiments with a large number of wafers of different material parameters indicate that under normal laboratory conditions, LIP can be observed only in some samples—samples that have moderate minority-carrier lifetime. We explain this observation and also show that wafer cleaning plays an extremely important role in passivation.

INTRODUCTION

It is well recognized that iodine-ethanol (I-E) or iodine-methanol (I-M) solutions can passivate Si surfaces [1–9]. This method of surface passivation can be very valuable when a temporary and removable passivation of a Si wafer surface(s) is needed. The most common application of a temporary passivation is for making lifetime measurements of Si wafers, where a very high-quality surface passivation is needed to measure the bulk lifetime, τ_b. This is a preferred method of passivation because, unlike other techniques of passivation such as deposition of nitride, oxide, or an n/p junction, I-E passivation is achieved at room temperature without any high-temperature wafer processing. Unfortunately, the mechanism of passivation is not well understood. As a result, it is a common experience that measurements made with I-E (I-M) solution are not reproducible. Earlier studies to investigate sources of these variations have shown that measurement is influenced by wafer cleaning [3,7,10]. Furthermore, Refs. [3,7] have shown that multiple cleaning of wafers leads to an improved passivation. Recently, we developed a procedure for wafer cleaning that yields highly reproducible values of minority-carrier lifetime as measured by photo-conductance decay (PCD) or quasi-steady-state photo-conductance (QSSPC) techniques. This cleaning procedure involves cleaning the wafer in Piranha, followed by a low-temperature oxidation and a hydrofluoric (HF) dip to remove a thin layer from the surface [7].

Because our new procedure enabled reproducible measurement of lifetime, it led to the discovery of another source of variability in lifetime measurement—the influence of light-exposure on the passivation produced by I-E. We observed that surface passivation is greatly hastened if the Si wafer, immersed in I-E solution, is exposed to light [10]. Here, we present a mechanism of passivation by I-E solution, which also explains the results presented in this paper on the dependence of passivation on the light illumination. In particular, we explain a very intriguing feature as to why only some wafers exhibit light dependence of the measured lifetime on light exposure. We will first briefly describe our experiment approach, including the improved cleaning procedure that has allowed us to achieve repeatable results and concomitantly allowed us to identify the light dependence.

EXPERIMENTAL DETAILS

Surface passivation was determined indirectly through the measurement of the minority-carrier lifetime. It is known that measured lifetime (τ_m) of a Si wafer can be expressed as: $1/\tau_m = 1/\tau_b + SRV/2W$, where τ_b, SRV, and W are the bulk lifetime, surface recombination velocity, and wafer thickness, respectively. Thus, for a given wafer of lifetime τ_b, the measured lifetime depends on the SRV. Because passivation reduces SRV, the passivation effect is manifested as an increase in the measured lifetime. Figure 1 shows effective lifetime, τ_m, for wafers of three different bulk lifetimes (10 μs, 100 μs, and 1 ms) as a function of SRV. It is seen that that the SRV must be reduced to about 1 cm/s to be able to measure the real τ_b of a 1-ms-lifetime wafer. However, for a 10-μs wafer, the SRV on the order of 1000 cm/s is sufficient. Hence, as is generally done, we also used the measurement of τ_m as a parameter to derive the effectiveness of the passivation.

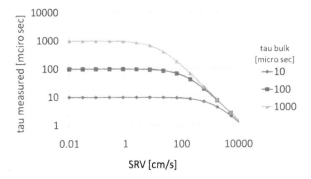

Figure 1. Dependence of measured lifetime, τ_m, on the actual bulk lifetime, τ_b, and surface recombination velocity, SRV.

This work was done primarily using single-side polished, single-crystal wafers within a large resistivity range of 1 Ω-cm to undoped. We used Czochralski (CZ) and float-zone (FZ) grown, both n-type and p-type wafers. The resistivity of each wafer was measured by four-point probe, and wafers were sorted into tight resistivity groups. Samples were cleaned using our recently developed cleaning procedure, placed in a polyethylene bag, and a few drops of I-E solution were placed on each side of the wafer. The bag was then allowed to collapse on the wafer and excess I-E was gently squeezed out. The wafer was then placed in a Sinton apparatus for measurement of τ_m. The Sinton machine allows different modes of measurement, which

138

include QSSPC, transient PCD, and generalized PC. When calibrated properly, we found that in generalized mode the injection-level dependence of τ_m was the same as measured by other modes. In view of this, we only describe the results measured by generalized PC. Passivation was determined simply on the basis of obtaining the highest lifetime value.

WAFER CLEANING

We (and other researchers) [3,7,10] have shown that wafer cleaning plays a very important role in obtaining good surface passivation. We have determined that even in a very high-quality wafer it is necessary to remove a thin surface layer to get good passivation. We believe that this need arises because some of the impurities diffuse several tens of Å into the surface and that in the presence of impurities it is difficult to get surface passivation that is needed to measure lifetimes > 1 ms. Although one can chemically etch a thin layer from a wafer surface, we found that a low-temperature optical oxidation, followed by an HF dip, can be very useful in removing a controlled layer from the surface. One important question is how deep a surface region must be removed or etched. To answer this question, we performed oxidation in steps to remove only about 75 Å in each step. The sample was cleaned and oxidized before each set of measurements. Figure 2 depicts the time dependence of the lifetime measurements for the first three cleaning steps.

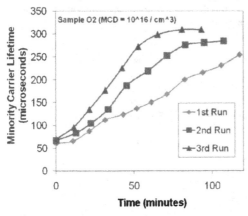

Figure 2. Time-dependence of τ_B after including oxidation in the cleaning procedure, for sequential cleaning steps. Each oxidation run removed a surface layer that was about 75 Å thick.

Figure 2 shows that surface cleanliness and its passivation are a function of total oxidation time. Based on this, we developed our procedure for wafer cleaning that consists of the following: (i) removal of organics by solvent cleaning followed by deionized (DI) water rinse, (ii) piranha (H_2SO_4:H_2O_2 2:1) cleaning at 80°C, (iii) dilute HF rinse, DI water rinse, and nitrogen drying, (iv) oxidation in an optical furnace for 10 min, and (v) dilute HF dip and N_2

139

drying. Following this cleaning, the samples were placed in a polyethylene bag. We found that zip-lock bags provided an excellent way to passivate the sample. We tried a variety of bags of different qualities and thicknesses, and the most convenient was a 1–2-mil-thick polyethylene bag. A well-cleaned sample was placed in a polyethylene bag and covered on both sides with I-E solution (typically of 0.1 molarity). Excess solution from each surface was squeezed out to leave a thin, uniform layer of the solution on the surface. In our measurements, the change in molarity of the solution between 0.01 and 0.1 did not influence the results. This cleaning-passivating procedure for lifetime measurement works very well and obtains highly reproducible results.

LIGHT-ACTIVATED PASSIVATION

Another feature of Figure 2 is that in each measurement it takes a long time before τ_{max} (maximum value of τ_m) is reached. We also observed that if the measurements were done at shorter intervals, the slope of the curves increased. We concluded that the light from the tester itself was influencing the measurement by lowering the surface recombination velocity and concomitantly causing the τ_m to increase. This interesting phenomenon indicates that the I-E surface passivation has a light-activated component. To confirm the light-induced passivation, we cleaned wafers (using our new oxidation procedure), placed them in an I-E bag, and exposed them to about 0.5-sun intensity from a solar simulator for 15 minutes. We found that the lifetime tester yielded τ_{max} immediately after the exposure; furthermore, there was a slow decrease (as shown in Figure 3a). We found that this decrease occurs for all wafers after the measured lifetime reached its maximum (see Figures 3a and 3b).

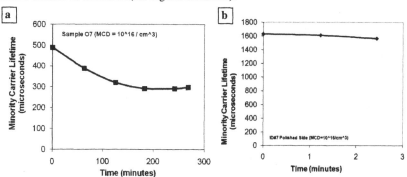

Figure 3. τ_m decay after light exposure in solar simulator (a); the initial increase was not observed. Short-term variation of τ_m for a long-lifetime wafer, (b).

Clearly, the light activation of passivation must be related to the chemistry of I-E. There is some published work [11–13] that proposes that the passivation of the Si surface is caused by tying up of silicon dangling bonds by the ethoxyl group, as illustrated in Figure 4. It has been proposed that this reaction is prompted by the dissociation of iodine.

140

Figure 4. Silicon surface passivated with ethoxylated group.

We first wanted to verify this assumption. To do so, we prepared a set of samples and placed them in a solution containing ethanol only and measured the lifetime. Then these samples were illuminated with light. Next, iodine was introduced, followed by exposure to light from a solar simulator. Figures 5a and 5b show typical results of the two groups. Both samples O3 (which exhibits LIP) and sample X14 (with no LIP effect) show that ethanol alone does not produce any passivation, even with illumination. It is clear that iodine is needed to produce passivation.

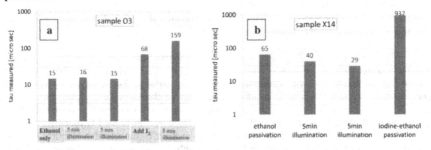

Figure 5. Effect of I_2 and illumination on a medium (a) and high (b) lifetime wafers.

Given that iodine is needed to produce passivation, it is possible that the ultraviolet (UV) component of light can cause dissociation of I_2, and the atomic iodine can further promote formation of Si-ethoxylate bonding. This assumption was tested on several wafers by placing each cleaned wafer in the I-E bag and then exposing it to UV light. However, the lifetime did not reach τ_{max}, as seen in the results of Figure 6. We also tried to determine if the effect was thermally induced because exposure of the wafer in the I-E bag caused it to heat up. Again, heating did not produce any change in the lifetime. Figure 6 elucidates the influence of various treatments on time dependence of the lifetime measured immediately after the treatment.

From the above results it is clear that surface passivation requires the presence of iodine. However, the results of Figure 6 show that iodine in methanol does not appear to dissociate rapidly by UV light, or perhaps there is a need to use higher light intensity. Only some wafers exhibit the effect of light in expediting the passivation process (as described), which suggests that passivation is related to some wafer parameter(s). We have attempted to relate LIP to material properties such as resistivity, oxygen content, carbon concentration, and other impurity levels. In this endeavor, we observed that the LIP effect was displayed by wafers that had τ_b in the range of about 50 to 400 ms (medium-lifetime group). High-lifetime and low-lifetime wafers

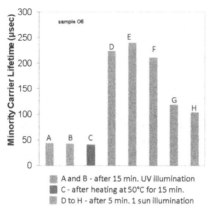

Figure 6. τ_m of a p-type wafer, resistivity 27 Ω-cm measured after several treatments: UV for 15 min and 30 min, heating, exposure to light for different times.

The mechanism of LIP appears to be related to carrier-induced dissociation of I_2. When a wafer in contact with I-E solution is illuminated with light of a broad spectrum, there is a large increase in the minority-carrier concentration at the Si surface. For a p-type wafer, a high concentration of electrons will be available and some of them can be injected into the I-E solution, where they dissociate I_2 producing Γ species. The active Γ can participate in converting a H-terminated surface (produced by HF dip performed just prior to I-E immersion) to an ethoxylate-terminated surface. The kinetics of passivation is illustrated in Figure 7. This reaction occurs in two phases. In phase 1, iodine simply acts as an oxidant. Because iodine is electronegative in nature, it has the ability to take in an extra electron to form ionized atomic iodine, Γ. Thus, when photogenerated electrons are available at the Si-H surface, I_2 can form iodine anions (Γ). The electron flow to iodine leaves Si-H in positively charged $[Si-H]^+$ ionic form. In this situation, an ethoxyl nucleophile can easily attach itself to Si-H via a coordinate bond, which occurs in the second phase. This occurrence is accompanied by release of hydrogen as a proton from the alcohol, leading to the charge neutrality of the five-coordinate Si species. Formation of five-coordinated Si in aqueous solutions is well known [11–13]. Next, the five-coordinated Si loses an electron, which is captured by iodine, leading to a positively charged five-coordinated Si. In the final step, the loss of another proton results in the formation of the final ethoxylated silicon surface.

The carrier-induced dissociation is likely to depend on the carrier concentration at the surface. The electron density at the surface depends strongly on the bulk lifetime of the wafer and its surface characteristics such as SRV. It is expected that carrier concentration at the surface of a long-lifetime wafer can be orders of magnitude higher than at the surface of a short-lifetime wafer. Hence, a long-lifetime wafer can get passivated even with low-intensity illumination, such as from room light. On the other hand, lower-lifetime material will require longer exposure to light. This explains why in our experiments that long-lifetime wafers, when properly processed, do not need simulator light for passivation. We believe that the same

142

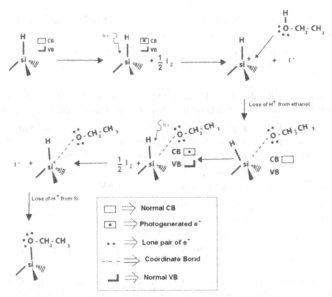

Figure 7. Proposed mechanism for passivation of silicon surface by I-E solution.

CONCLUSIONS

We presented experimental results showing that illumination of a silicon sample embedded in I-E solution causes optically induced enhancement of surface passivation. The mechanism of passivation appears to be related to dissociation of I_2 into I^- ions by the photo-generated carriers. We have proposed a detailed reaction outlining the participation of iodine and photo-generated electrons in producing a passivated Si surface terminated by ethoxylate groups. Our experiments indicate that generally good passivation of silicon wafers requires two essential steps: (i) wafer cleaning, which includes removal of about 200–300 Å of Si from each surface, and (ii) exposure of an I-E-passivated wafer to spectrum-rich light. We outlined a procedure that yields a very clean surface and have found that using fresh chemicals (piranha, HF, and other acids) for each batch of wafers minimizes surface-quality variations. Our experience is that these chemicals have the propensity to acquire impurities from ambient and, in some cases, can leach them from the containers if very high-quality containers are not used. We suggest using optical oxidation after piranha cleaning.

ACKNOWLEDGEMENTS

This work was supported by the U.S. Department of Energy under Contract No. DE-AC36-08GO28308 with the National Renewable Energy Laboratory.

REFERENCES

1. A. G. Aberle, "Surface passivation of crystalline silicon solar cells: A review," *Progress in Photovoltaics,* vol. **8**, pp. 473–487, Sep–Oct 2000.

2. T. Maekawa and Y. Shima, "Effect of steady bias light on carrier lifetime in silicon wafers with chemically passivated surfaces," *Japanese Journal of Applied Physics Part 2 - Letters,* vol. **35**, pp. L133–L135, Feb 1 1996.

3. H. Msaad, J. Michel, A. Reddy, and L. C. Kimerling, "Monitoring and optimization of silicon surface quality," *Journal of the Electrochemical Society,* vol. **142**, pp. 2833–2835, Aug 1995.

4. M. R. Page, Q. Wang, T. H. Wang, Y. Yan, S. W. Johnston, and T. F. Ciszek, "Silicon surface and heterojunction interface passivation studies by lifetime measurement," in *13th Workshop on Crystalline Silicon Solar Cell Materials and Processes* Vail, Colorado, 2003.

5. T. M. Renee et al., "Atomic-scale mechanistic study of iodine/alcohol passivated Si(100)," presented at *196th Meeting of the Electrochemical Society,* 1999.

6. N. A. Zarkevich and D. D. Johnson, "Energy scaling and surface patterning of halogen-terminated Si(001) surfaces," *Surface Science,* vol. **591**, pp. L292–L298, Oct 20 2005.

7. B. L. Sopori, P. Rupnowski, J. Appel, V. Mehta, C. Li, and S. Johnston, "Wafer preparation and iodine-ethanol passivation procedure for reproducible minority-carrier lifetime measurement," in *33rd Photovoltaic Specialists Conference* San Diego, CA, 2008.

8. T. S. Horányi, T. Pavelka, and P. Tüttö, "In situ bulk lifetime measurement on silicon with a chemically passivated surface," *Applied Surface Science,* vol. **63**, pp. 306–311, 1993.

9. J. R. William, J. M. David, and S. L. Nathan, "Role of inversion layer formation in producing low effective surface recombination velocities at Si/liquid contacts," *Applied Physics Letters,* vol. **77**, pp. 2566–2568, 2000.

10. G. J. Norga, M. Platero, K. A. Black, A. J. Reddy, J. Michel, and L. C. Kimerling, "Detection of metallic contaminants on silicon by surface sensitive minority carrier lifetime measurements," *Journal of the Electrochemical Society,* vol. **145**, pp. 2602–2607, Jul 1998.

11. J. A. Haber and N. S. Lewis, "Infrared and X-ray photoelectron spectroscopic studies of the reactions of hydrogen-terminated crystalline Si(111) and Si(100) surfaces with Br-2, I-2, and ferrocenium in alcohol solvents," *Journal of Physical Chemistry B,* vol. **106**, pp. 3639–3656, Apr 11 2002.

12. D. J. Michalak, S. Rivillon, Y. J. Chabal, A. Esteve, and N. S. Lewis, "Infrared spectroscopic investigation of the reaction of hydrogen-terminated, (111)-oriented, silicon surfaces with liquid methanol," *Journal of Physical Chemistry B,* vol. **110**, pp. 20426–20434, Oct 2006.

13. S. D. Solares, D. J. Michalak, W. A. Goddard, and N. S. Lewis, "Theoretical investigation of the structure and coverage of the Si(111)-OCH3 surface," *Journal of Physical Chemistry B,* vol. **110**, pp. 8171–8175, Apr 2006.

Mater. Res. Soc. Symp. Proc. Vol. 1123 © 2009 Materials Research Society   1123-P02-02

Infrared ellipsometric characterization
of silicon nitride films on textured Si photovoltaic cells

M. F. Saenger[1], M. Schädel[3], T. Hofmann[1], J. Hilfiker[2], J. Sun[2], T. Tiwald[2], M. Schubert[1] and J. A. Woollam[1,2]

[1]Department of Electrical Engineering, and Nebraska Center for Materials and Nanoscience, University of Nebraska-Lincoln, Lincoln, NE, 68588-0511, USA
[2]J. A. Woollam Co. Inc., Lincoln, NE, 68508-2243, USA
[3]Q-Cells AG, Thalheim, 06766, Germany

ABSTRACT

We present an infrared spectroscopic ellipsometry investigation of Si_xN_y films deposited on textured Si substrates employed for photovoltaic cells. A multiple-sample data analysis scheme is used in order to determine the Si_xN_y dielectric function and thickness parameters regardless of the surface morphology of the substrate. We observe changes in the dielectric function of the silicon nitride film which suggest variations in the chemical composition of the films depending on the substrate morphology.

INTRODUCTION

Surface texturing is a commonly used technique used to reduce reflection losses and thus increase the quantum efficiency of silicon solar cells. Typical texturing methods include mechanical processes or chemical reactions involving alkaline or acidic etchants producing different surface morphologies [1-4]. In order to further suppress reflection losses, an antireflecting (AR) thin film is commonly added to the textured surface, making AR films an important feature in a high efficiency solar cell design. Si_xN_y is a common material used for the AR film on silicon solar cells due to its passivation, adhesive, and optical properties. The film thickness together with the optical constants determine how effectively the reflected light is suppressed for an optimal design wavelength, usually around 550 nm for photovoltaic applications. Due to the striven low reflectance and high scattering, the optical characterization of the thin film thickness and optical constants parameters renders as a difficult task, such that advanced techniques or measurement geometries are necessary to analyze them [5]. Previous investigations in the near-infrared to ultraviolet spectral range showed that the optical constants of the Si_xN_y films on textured silicon wafers can be well approximated by an effective medium approximation (EMA) that combines the optical constants of a reference film with a void fraction which varies depending on the substrate morphology [5]. However, it is not clear if the changes have a pure physical origin or if they are also due to variations in the films chemistry. Investigations in the infrared (IR) spectral range provide an option to overcome the low reflection problem and also access to the film thickness parameter values. Furthermore, it is possible to study chemical variations in the film by analyzing the corresponding phonons or chemical bond signatures of silicon nitride detectable in the IR spectral range. The vibration frequencies for the $N-Si_3$, Si-N, Si-O-Si and N-H modes are in the range between 300 and 1200 cm^{-1} and thus their strength and resonance frequency should be measurable with infrared spectroscopic ellipsometry (IRSE). IRSE has been proven as an efficient technique to obtain information about phonons, chemical composition and thickness parameters, which makes it the tool of choice for the present investigation [6]. Ellipsometry determines the complex reflectivity ratio ρ. In general, ellipsometry measurements are typically reported in terms of the parameters Ψ and Δ, which are related to the magnitude and phase of the complex reflectivity ratio as defined by $\rho \equiv \tan(\Psi)e^{i\Delta} = R_p / R_s$, where R_p and R_s denote the complex Fresnel reflection coefficients [7, 8]. Here we report on spectroscopic ellipsometry investigations of the dielectric function (ε) in the IR spectral range (300 – 1200 cm^{-1}) of silicon nitride films on differently textured silicon substrates for photovoltaic cell applications. From our model analysis we find differences in the strength of the polarity of lattice vibrations, which suggest a variation in the film chemical composition among the differently textured substrates. The thickness parameter values determined from the IR data analysis are in good agreement with previous ellipsometry analysis in the near-IR to ultraviolet (NIR-UV) spectral range [5].

EXPERIMENTAL

The wafers employed as substrate consisted on a polished silicon wafer (reference), an alkaline etched multi- and monocrystalline (m-Alkaline and c-Alkaline) wafers, a multicrystalline acidic etched wafer (m-Acid) and a multicrystalline "as-cut" (m-Ascut) wafer. The samples studied in this work have been previously investigated, further details regarding the samples preparation and morphology can be found in the literature [5]. The Si_xN_y films under investigation were deposited with a commercially available plasma-enhanced chemical vapor deposition (PECVD) system. On each wafer, four equal sized spots of ca. 3.5 cm diameter were coated with a Si_xN_y film under identical deposition conditions. The film thickness of the four spots on each wafer varied in deposition steps of approximately 70 nm such that the film thicknesses was approximately 70, 140, 210 and 340 nm. For the ellipsometric measurements in the IR spectral range a Fourier-transform-based spectroscopic ellipsometer was employed. The measurements were performed in the spectral range from 300 to 5000 cm^{-1} at angles of incidence of $\Phi = 55°$ and 70°.

DISCUSSION

Figure 1 shows the experimental, the effective medium approximation (EMA) and the best-match data for the ellipsometric parameters Ψ and Δ in the IR spectral range. The EMA approach is explained further below. The ellipsometric data was analyzed using a stratified model calculation [7], which consisted of two layers [substrate / Si_xN_y film]. Only for the reference wafer an intermediate layer was included which accounted for the native silicon dioxide.The thickness of the silicon dioxide was measured on the uncoated area and the thickness was of approx. 2 nm, which is a typical value for native oxide in silicon. This value (2 nm) was used for the models of the films and was not further varied during the data analysis. The optical constants of the uncoated textured substrates were obtained through a direct point-by-point (pbp) analysis, which consists of a direct conversion of the ellipsometric data into optical constants, as previously demonstrated [9]. The optical constants for SiO_2 and Si used for the reference wafer data analysis can be found in the literature [10]. The dielectric function of the Si_xN_y films was studied using the "multiple-sample" data analysis method [9], which involved the simultaneous analysis of the measurements of the four films of different thicknesses on each textured substrate. The dielectric function of the Si_xN_y layer was described by a parametric model

146

composed by a set of Gaussian oscillators, as employed previously [12-14]. From Fig. 1 we can observe the good match between the experimental and the parametric model data which offers the best-match model. The parametric model approach is described in detail elsewhere [7, 8]. The oscillator parameterization describes both, the real and the imaginary parts of the dielectric function in the studied IR spectral region and accounts for specific chemical bonds in Si_xN_y.

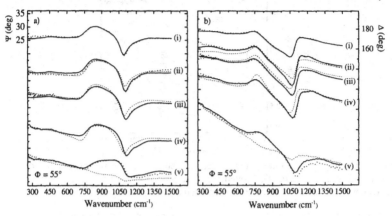

Figure 1: Experimental (dotted), effective medium approximation (dashed), and best match calculation (solid line) spectra of the ellipsometric parameter Ψ (a) and Δ (b), respectively, measured at an angle of incidence of $\Phi = 55°$. The data corresponds to the Si_xN_y layer of nominal thickness of approx. 210 nm on the different textured wafers: i) reference, ii) m-Alkaline, iii) m-Ascut, iv) m-Acid and v) c-Alkaline. The Ψ and Δ spectra are shifted in steps of 10° with respect to the reference spectra for a better visualization.

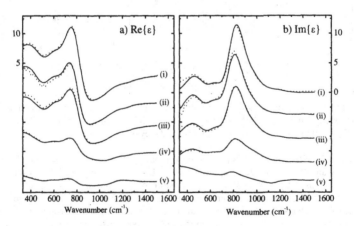

Figure 2: Real (a) and imaginary (b) parts of the point-by-point extracted (dotted) and best-match calculation (solid lines) dielectric function of the silicon nitride films for the different substrate types, as described in Fig. 1. The Re{ε} (Im{ε}) data are shifted in steps of 5 (3.5) for a better visualization.

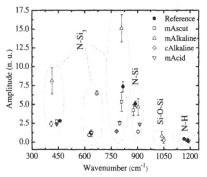

Figure 3: Oscillator amplitude versus resonance energy model parameters for the different textured wafers studied here. The dotted areas are plotted as a guide for the eyes to denote the assigned chemical bond of the corresponding oscillators.

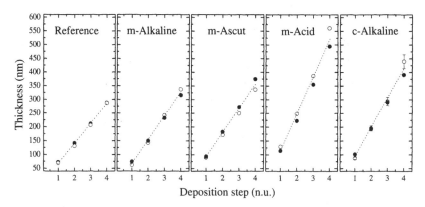

Figure 4: Thickness parameter values plotted versus deposition steps for the different textured wafers obtained from the UV-VIS (solid symbols [5]) and IR (empty symbols) ellipsometry data analysis. The dotted lines are plotted as guides for the eyes.

In analogy to the NIR-UV analysis [5], a second model for the silicon nitride dielectric function consisting of an effective medium approximation (EMA) was employed for comparison. The EMA was composed by a Bruggeman type mixture of the parametric model obtained optical constants of the Si_xN_y film on the reference wafer and a void fraction which was allowed to vary during the data analysis. While the EMA can be employed as a good approximation in the NIR-UV spectral range, in the IR spectral range the EMA model provides a very poor description of

148

the experimental data compared to the Gaussian oscillators model, as can be seen from Fig. 1. However, in the NIR to UV analysis performed in the same samples [5], the EMA was successful. This apparent contradiction appears because the IR spectral range is also sensitive to phonon and chemical bond modes while in the NIR-UV spectral range no modes exist in the studied films and the dispersion mostly depends on the higher energy electronic transitions.

Figure 2 shows the good match between the point-by-point and the parametric model obtained dielectric function of the silicon nitride films on the different substrates. The spectra are arranged from top to bottom with increasing void fraction as found from the EMA analysis in the NIR-UV [5]. From this figure we can see that the amplitude of the spectra decreases and that the line shape also varies from substrate to substrate. This can be explained by two effects, the texturization of the substrate which causes an apparent decrease of the effective optical constants, and by a change in the distribution of the strength and resonance energy of the chemical bonds which is the reason why the EMA does not provide a good description of the IRSE ellipsometry data.

Figure 3 shows the Gaussian oscillator parameters for the amplitude plotted versus the oscillator resonance energy parameter. Of special interest is the spectral range below 1200 cm^{-1} which is where the signature of most chemical bonds is located. The mode assignation is based on reports on the infrared absorption of silicon nitride films [12-14]. The most intense peaks in this region are due to: N–H bending (1170 cm^{-1}), -Si–O–Si- asymmetrical stretching (1050 cm^{-1}), and two Si–N stretching (960 and 840 cm^{-1}) vibrational modes [16-19]. According to the literature the three absorption bands (near 880, 640, and 480 cm^{-1}) are due to a vibration of the N-Si$_3$ bonding unit [10]. We observe a variation of the different chemical bonds for the different textured wafers.

Figure 4 shows the thickness parameters for the films on the different wafer textures obtained from the IR and from the NIR-UV analysis [5]. The IR data is consistent with the data from the NIR-UV analysis. The origin of the differences is assumed to be due to thickness non-uniformities in the sample. The variations between the NIR-UV and IR analysis obtained values of the film thickness parameters are relatively small and are attributed to thickness non-uniformities in the films. We observe in Fig. 4 that the thickness values are always bigger for the textured wafers compared to the reference wafer. This observation together with the obtained different strength of the bonds seen in Fig. 3, suggests that the growth rate of the films is affected by a variation on the chemistry caused by the substrate previous treatment or micro-morphology.

CONCLUSIONS

The effect of the texture on the dielectric function causes an apparent decrease of the dielectric function. In the IR spectral range we observe that in addition to the lowering of the optical constants the dielectric function differs for each substrate texture which causes the EMA model to fail. The observed decrease of the dielectric function is in good agreement with previous analysis in the NIR-UV spectral range and furthermore we are able to detect variations in the dielectric function due to variations in the chemical stoichiometry of the films from substrate to substrate. We found for the m-Alkaline sample an additional mode in the IR spectra related to a Si-O-Si vibration suggesting a different composition to the other films. We found a strong variation in the amplitudes of the chemical bonds between 300 and 1500 cm^{-1}, further studies in dependence of the chemical composition should follow.

149

ACKNOWLEDGEMENTS

We (M.Sch.) gratefully acknowledge support by the NSF within MRSEC Q-SPINS, by the J. A. Woollam Foundation, and by startup funds from The College of Engineering of the University of Nebraska-Lincoln.

REFERENCES

1. C. R. Baraona, and H. W. Brandhorst, Proceedings of the Eleventh IEEE Photovoltaic Specialists Conference (Institute of Electrical and Electronits Engineers, New York) 44 (1975).
2. M. A. Green, Prog. Photovolt: Res. Appl. 7, 317 (1999).
3. S. de Wolf, P. Choulat, E. Vazsonyi, R. Einhaus, E. v. Kerschaver, K. de Clercq, and J. Szlufcik, in Proc. 16th EPVSEC, Glasgow, 1521 (2000).
4. J. D. Hylton, A. R. Burgers, and W. C. Sinke, J. Electrochem. Soc., 151, G408 (2004).
5. Spectroscopic ellipsometry characterization of SiyNx antireflection films on textured multicrystalline and monocrystalline silicon solar cells, M. F. Saenger, J. Sun, M. Schädel, J. Hilfiker, M. Schubert, and J. A. Woollam, Thin Solid Films, (in press).
6. A. Kasic, M. Schubert, T. Frey, U. Köhler, D. J. As, and C. M. Herzinger, Phys. Rev. B 65, 184302 (2002).
7. H. G. Tompkins and E. A. Irene, eds., *Handbook of Ellipsometry* (William Andrew Publishing, Highland Mills, 2004).
8. R. M. A. Azzam and N. M. Bashara, *Ellipsometry and Polarized Light* (North-Holland Publ. Co., Amsterdam, 1984).
9. T. Hofmann, G. Leibiger, V. Gottschalch, I. Pietzonka, and M. Schubert, Phys. Rev. B 64, 155206 (2001).
10. H. R. Philipp, *Handbook of Optical Constants in Solids*, edited by E. D. Palik (Academic, Boston, 1985), p. 749.
11. C. M. Herzinger, B. Johs, W. A. McGahan, J. A.Woollam, and W. Paulson, J. Appl. Phys. 83 (6), 3323 (1998).
12. Z. Yin and F. W. Smith, Phys. Rev. B 42, 3666 (1990).
13. J. J. Mei, H, Chen, W. Z. Shen and H. F. W. Dekkers, J. Appl. Phys. 100, 073516 (2006).
14. M. Klanjšek Gunde and M. Maček, Phys. Status Solidi A 183 (2), 439 (2001).
15. J. Humlíček, R. Henn, and M. Cardona, Phys. Rev. B 61, 14554 (2000).
16. D. V. Tsu, G. Lucovsky, and M. J. Mantini, Phys. Rev. B 33, 7069 (1986).
17. Jiun-Lin Yeh and Si-Chen Lee, J. Appl. Phys. 79, 656 (1996).
18. H. Ono, T. Ikarashi, Y. Miura, E. Hasegawa, K. Ando, and T. Kitano, Appl. Phys. Lett. 74, 203 (1999).
19. B. Pivac, J. Mater. Sci. Lett. 12, 23 (1993).

Mater. Res. Soc. Symp. Proc. Vol. 1123 © 2009 Materials Research Society   1123-P03-08

Evaluation of Four Imaging Techniques for the Electrical Characterization of Solar Cells

Gregory M. Berman[1,3], Nathan J. Call[2,3], Richard K. Ahrenkiel[2,3], and Steven W. Johnston[3]

[1]Department of Electrical Engineering, University of Colorado, Boulder, CO 80309, U.S.A.
[2]Department of Materials Science, Colorado School of Mines, Golden, CO 80401, U.S.A.
[3]National Renewable Energy Laboratory, Golden, CO 80401, U.S.A.

ABSTRACT

We evaluate four techniques that image minority carrier lifetime, carrier diffusion length, and shunting in solar cells. The techniques include photoluminescence imaging, carrier density imaging, electroluminescence imaging, and dark lock-in thermography shunt detection. We compare these techniques to current industry standards and show how they can yield similar results with higher resolution and in less time.

INTRODUCTION

In this paper, we address four imaging techniques that measure minority carrier lifetime, diffusion length, and the location of shunts in a solar cell, and we compare our techniques to methods currently used in industry. Minority-carrier-lifetime and diffusion-length measurements are critical for characterizing the quality of solar materials to predict the efficiency of a device made from them [1-2]. Shunts can drastically lower a solar cell's efficiency and are introduced in the processing of a wafer into a working solar cell [3].

Minority carrier lifetime is the average time it takes for a free carrier generated in a material to recombine. We measure it with photoluminescence imaging (PLI) and carrier-density imaging (CDI). PLI probes the radiative recombination of an optically excited sample, which is proportional to the sample's carrier density [4-6]. CDI measures the transmission of infrared (IR) light through an optically excited sample, which is also proportional to carrier density [7-8]. We compare these two carrier-lifetime techniques to each other and to microwave-reflection photoconductive decay on a Semilab tool, an industry standard that relates a material's conductivity to its carrier density.

Carrier diffusion length is the average distance a free carrier travels in a material before it recombines. We measure it with electroluminescence imagining (ELI). ELI probes the radiative recombination of a finished solar cell that is electrically driven in forward bias [9-10]. We compare ELI images with light-beam-induced current (LBIC) maps [11], an industry standard that measures the optically induced current in a solar cell.

Shunts are defects created during solar cell processing that leak current. We use dark lock-in thermography (DLIT) to detect them. While shunts can be detected by measuring a cell's I-V (current-voltage) curve, no information about a shunt's cause or location is identified. In DLIT imaging, the solar cell is put under reverse bias, and the shunts are detected by a thermal signature, generated by current leaking across the p-n junction [12]. We demonstrate shunt detection and identify the defect with a scanning electron microscope.

METHODS

Photoluminescence imaging

PLI is the steady-state equivalent of time-resolved photoluminescence that is imaged onto a camera. This technique has recently been made possible by the high sensitivity and large pixel counts of modern Si charge-coupled-device (CCD) cameras. During PLI, carriers in a wafer or a finished cell are optically excited. The camera then detects a small amount of radiative recombination from the indirect bandgap material. Areas on the sample that have long minority-carrier lifetimes will have a higher steady-state density of free carriers, resulting in a stronger radiative recombination signal.

We collect PLI data using a PIXIS 1024BR Si CCD camera (Princeton Instruments/Acton). This camera has a 1024 x 1024 array of 13 μm pixels and is back illuminated and cooled to 200 K. The detector is deep depleted to enhance quantum efficiency beyond 1100 nm. A compact lens (Schneider Optics Cinegon) is mounted to the camera along with a stack of two 810 nm notch filters (Kaiser Optical Systems). A black glass filter (Schott RG1000) is placed between the two notch filters to eliminate resonance. The filter stack is designed to sufficiently attenuate reflected light from the 60 W, 810 nm laser diode excitation source. The fiber output from the laser diode is expanded to the sample area by a collimator and an engineered diffuser.

Measurements are made while the sample is under steady-state illumination. The image is a summation of all photons collected in a set exposure time. It is a spatial map of the radiative recombination in the sample, which is directly related to carrier concentration and carrier lifetime. While such images are qualitative, we show that they can be calibrated by transient-lifetime measurements to yield absolute lifetime values.

Carrier density imaging

CDI is a direct measurement of free carriers in a process called free-carrier absorption. Free-carrier absorption occurs when excited carriers decrease the transmission of mid-to-far infrared light in a semiconductor. During this process, a wafer without any metallization is optically excited. Then an infrared camera images the change in transmission of black-body radiation through a sample with and without the excited carriers.

CDI data is acquired with a Silver 660M InSb infrared camera (ElectroPhysics/Cedip), which has a 640 x 512 array of 15 μm pixels and a spectral response from 3.6–5.1 μm. A built-in Stirling stage cools the detector to ~76 K. The camera has a built-in lock-in feature and a maximum frame rate of 100 Hz. We use the camera's lock-in mode to collect CDI data with the laser diode driven by a 27 Hz square wave. The lock-in mode greatly reduces the noise of the measurements — a critical factor for measuring the small signals in CDI. We used a hot plate set at ~400 K as the black-body source. To enhance the hotplate's emissivity, we have coated it with a black high-temperature paint.

The images we collect represent the amplitude of the Fourier transform of each pixel in time at the particular frequency we are collecting data. The images show the relative change in transmission spatially across the wafer. This transmission is proportional to the lifetime. While these images are qualitative by themselves, they can be calibrated by a single-point lifetime measurement to yield accurate lifetime images.

Electroluminescence imaging

We collect ELI data in much the same way as PLI. We use the same Si CCD camera, though we take off the filter stacks since there is no laser excitation light to filter out. Others have used various filter arrangements to enhance ELI data and analysis [10]. In ELI, a finished solar cell is electrically driven in forward bias like a light emitting diode. As in PLI, the measurement is taken under steady-state excitation with a typical exposure time of one second. Raw ELI data is qualitative, but can be made quantitative by the method of data collection and analysis [9].

Dark lock-in thermography

To detect shunts using DLIT imaging, a finished wafer is electrically driven in reverse bias. The reverse bias is applied to the sample by an amplifier driven by a square wave. An ideal solar cell under reverse bias would have very little current flow. However, imperfections in the cell often lead to areas where there is a lower virtual resistance through the cell's junction, causing current to flow through it. This flow of current then generates heat, creating hotspots that we image in lock-in mode with the same thermal camera used in CDI.

RESULTS AND DISCUSSION

Comparison of PLI and CDI to transient lifetime measurements

Our data show that there is a strong correlation between PLI and CDI as well as with the actual measured lifetime on a wafer. For the absolute measurement, we use the microwave–reflection–photoconductive-decay technique built into a Semilab lifetime-measurement tool. This technique measures the conductivity of a material, which is proportional to the carrier density, by detecting the change in reflection of microwaves off the sample. The lifetime is then found from the carrier-concentration decay rate after a laser pulse illuminates the sample. It then takes ~6 minutes to map a 5-inch standard wafer at a resolution of 1 mm. Resolution is limited to the excitation light cross-section incident on the sample. A PLI image of the same wafer is collected in ~1 second at a resolution of ~125 µm. Likewise, a CDI image is made in ~10 seconds with a resolution of ~250 µm. The resolution is ultimately limited by the diffraction limit of light, which is ~1 µm for PLI and ~5 µm for CDI. The disadvantage of these imaging techniques is that they do not yield quantitative results. The lack of quantitative results is shown in Fig. 1, where only the Semilab map has a scale for the lifetime.

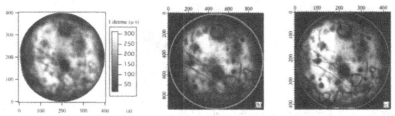

Figure 1. Lifetime of CZ Si wafer with induced impurities acquired from (a) a Semilab lifetime scanner, (b) photoluminescence imaging, and (c) carrier-density imaging.

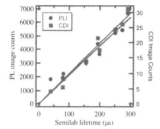

To obtain a quantitative analysis of the imaging techniques, we compare identical points on a wafer with each technique and plot them against each other. We find that there is a linear relationship between the pixel counts, the signal strength at a single pixel, and the measured lifetime values on the Semilab tool. By measuring two different points on a wafer with the Semilab tool, we can create a calibration constant to transform both PLI and CDI data into quantitative lifetime values, as shown in Fig. 2.

Figure 2. Comparison of PLI and CDI data with Semilab lifetime measurements at discrete points on a CZ Si wafer to calibrate image values to absolute lifetime.

Comparison of ELI to LBIC diffusion-length measurements

The standard method of mapping carrier-diffusion length in a finished cell is through the LBIC technique. Mapping a 5-inch square wafer takes ~6 hours at a resolution of 250 μm, as seen in Fig. 3(a) and (b). For the same wafer, ELI images can be collected in ~1 second at a resolution of ~125 μm, as seen in Fig. 4(a) and (b). Although the correlation between the LBIC measurement and ELI may not appear linear, Fig. 3(c) and 4(c) illustrate the similarity between them. Such correlations have been quantified in detail previously [9].

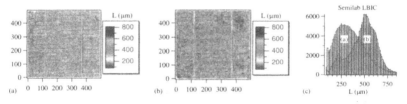

Figure 3. LBIC carrier-diffusion-length map for (a) a low-efficiency wafer and (b) a high-efficiency wafer. (c) Comparitive histogram of the diffusion lengths of (a) and (b).

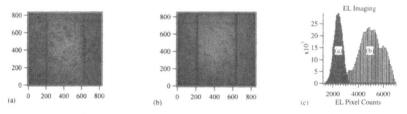

Figure 4. Electroluminescence imaging of the relative carrier-diffusion length for (a) a low-efficiency wafer and (b) a high-efficiency wafer. (c) Comparative histograms of the image pixel counts of (a) and (b).

Shunt Detection and Analysis using DLIT

Using DLIT, we can find the location of all the major shunts on a wafer and gain insight into their cause as shown in Fig. 5. We can determine which shunts are leaking the most current from the amplitude image [Fig. 5(b)]. In the phase image [Fig. 5(c)], we can see how fast heat is spreading from different points in the sample, which allows us to resolve smaller shunts.

Figure 5. Thermographic shunt detection in a multicrystaline Si wafer at 4 V reverse bias. (a) Thermal image. (b) Lock-in amplitude image. (c) Lock-in phase image. Units are arbitrary.

To get a better understanding of what may be causing shunts, we use a microscopic objective on our thermal camera to view specific shunts at a higher optical resolution. Often a shunt can be seen on a gridline where the metallization has punctured through the emitter layer. In some cases, a bright shunt in the shape of a line corresponds to a crack in the wafer that was filled in during metallization, causing a catastrophic shunt that ruins the device. For the case shown in Fig. 6, an aluminum particle has punctured through the emitter layer, creating a path for current to flow. In the zoomed-in DLIT image, the shunt seems asymmetric. A scanning electron microscope image [Fig. 6(b)] shows that this asymmetric shunt corresponds to an embedded aluminum particle.

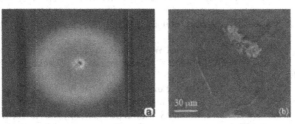

Figure 6. Images of a shunt caused by an embedded aluminum particle using (a) a thermal camera in lock-in mode with ~50X magnification and (b) a scanning electron microscope.

CONCLUSIONS

We have demonstrated imaging methods for minority carrier lifetime (PLI, CDI) and diffusion length (ELI) that give much faster and higher resolution data than traditional single-point mapping techniques. For carrier lifetime imaging, we have shown that a single-point calibration can easily transform a qualitative image into actual lifetime values. These imaging techniques enable the possibility of higher-level quality control and defect analysis of solar cell materials in in-line production processes.

155

We have used DLIT to find the location of shunts on a wafer and trace them back to specific points in the production process. In industry, this information could be used to address problems in production and increase the yield of working wafers with higher efficiencies.

ACKNOWLEDGMENTS

We thank CaliSolar Inc. for providing solar cells, Dr. Jian Li and Mr. Jerry Tynan for their assistance in the experimental set-up, Mr. Bobby To for taking the SEM images, and Dr. Mowafak Al-Jassim and Dr. Dean Levi for coordinating this work. This work was supported by the U.S. Department of Energy under Contract No. DE-AC36-08GO28308 with the National Renewable Energy Laboratory (NREL) and under NREL's Solar America Initiative PV Incubator program.

REFERENCES

1. D. K. Schroder, *Semiconductor Material and Device Characterization*, (John Wiley & Sons, Inc., New York, 1990).
2. J. W. Orton and P. Blood, *The Electrical Characterization of Semiconductors: Measurement of Minority Carrier Properties* (Academic Press, Inc., San Diego, CA 1990).
3. O. Breitenstein and M. Langenkamp, *Lock-in Thermography – Basics and Uses for Functional Diagnostics of Electrical Components* (Springer, New York, 2003).
4. Y. Koshka, S. Ostapenko, I. Tarasov, S. McHugo, and J. P. Kalejs, *Appl. Phys. Lett.* **74**, 1555 (1999).
5. M. Tajima, Z. Li, S. Sumie, H. Hashizume, and A. Ogura, *Jpn. J. Appl. Phys.* **43**, 432 (2004).
6. T. Trupke, R. A. Bardos, M. C. Schubert, and W. Warta, *Appl. Phys. Lett.* **89**, 044107 (2006).
7. M. Bail, J. Kentsch, R. Brendel, and M. Schulz, *Proceedings of the 28th IEEE-PVSC*, Anchorage, AK, 99 (2000).
8. J. Isenberg, S. Riepe, S. W. Glunz, and W. Warta, *J. Appl. Phys.* **93**, 4268 (2003).
9. T. Fuyuki, H. Kondo, T. Yamazaki, Y. Takahashi, and Y. Uraoka, *Appl. Phys. Lett.* **86**, 262108 (2005).
10. P. Wurfel, T. Trupke, M. Rudiger, T. Puzzer, E. Schaffer, W. Warta, and S. W. Glunz, *22nd European Photovoltaic Solar Energy Conference and Exhibition*, Milan, Italy (2007).
11. O. Palais, J. Gervais, E. Yakimov, and S. Martinuzzi, Eur. Phys. J. AP **10**, 157 (2000).
12. O. Breitenstein in *17th Workshop on Crystalline Silicon Solar Cells and Modules: Materials and Processes*, edited by B. L. Sopori (2007).

Mater. Res. Soc. Symp. Proc. Vol. 1123 © 2009 Materials Research Society   1123-P01-06

SIMS Study of C, O and N Impurity Contamination for Multi-Crystalline Si Solar Cells

Larry Wang and R. S. Hockett
Evans Analytical Group, 810 Kifer Road, Sunnyvale, CA 94086, U.S.A.

ABSTRACT

This paper is a case study of using SIMS to quantitatively measure C, O and N impurity contamination at two sequential *commercial* process steps: (1) Si feedstock: 7N (modified Siemens) and 5N feedstock (UMG-Si); and (2) multi-crystalline (mc-Si) solar wafers: cut and etched, from directional solidification bricks grown from 7N and 7N/5N (80:20) feedstock. The conclusion of this study is twofold: (a) the primary opportunity to reduce C, O and N contamination in mc-Si solar cells is at the directional solidification process, and (b) the costly specification of highly pure Si feedstock is unnecessary *from a C, O and N perspective* if a directional solidification process is used.

INTRODUCTION

Oxygen, carbon and nitrogen are important elements in silicon photovoltaic (PV) technology.

Oxygen

Oxygen forms a BO_x defect in p-type Si wafers which are used in most Si solar cells, and this defect is responsible for the initial light-induced degradation of solar cell efficiency [1]. Controlling the oxygen concentration below some prescribed level is one approach to reduce this efficiency loss. The SEMI PV wafer standard M6-0707 [2] prescribes that the oxygen concentration should be below $8 \times 10^{17}/cm^3$ in multi-crystalline wafers and $1 \times 10^{18}/cm^3$ in single crystal wafers.

Oxygen can enter the silicon from a variety of paths, such as, (1) the crucible material used in Cz-pullers and in directional solidification furnaces; (2) the process of making Si feedstock bricks if it some kind of oxide liner is used in the brick formation; or (3) the process for upgrading metallurgical-grade silicon (UMG-Si) as feedstock for Si solar wafers.

The measurement of oxygen in silicon at levels of $8 \times 10^{17}/cm^3$ implies the detection limit should be below $8 \times 10^{16}/cm^3$. The chosen analytical method can depend upon the form of the silicon to be tested. For example, FTIR can be used for single crystal silicon wafers according to SEMI MF1188-1107 [3]. However, this method has some special challenges for multi-crystalline wafers in that there may be considerable oxygen precipitation which is not quantified by FTIR, and the backside IR reflectance is most likely not matched to the BLANK silicon wafer used in the method, thus giving a bias error the determination of the oxygen content. If the oxygen needs to be measured in feedstock bricks, chunks, granules or flakes, FTIR is not appropriate.

SIMS can measure the oxygen concentration in all these forms of silicon. SIMS has been used for over 20 years to measure [O] in heavily-doped single crystal electronic-grade silicon according to SEMI MF1366-1107 [4], and a variant of this method can be used for non-wafer physical forms as well.

Carbon

Carbon can form SiC-type defects including filaments in multi-crystalline PV Si. These defects can cause local shunting behavior [5] and may affect minority carrier lifetime. But another reason to limit the carbon concentration is the formation of rod-like SiC defects in multi-crystalline PV blocks that can break wire saws used to slice the wafers. The SEMI PV wafer standard prescribes that carbon concentration should be below $1 \times 10^{18}/cm^3$ in multi-crystalline wafers and $5 \times 10^{17}/cm^3$ in single crystal wafers.

Carbon can enter the silicon from a variety of paths, such as, (1) the crucible material used in Cz-pullers or other furnaces; (2) from the open ambient in directional solidification furnaces; (3) the process of making feedstock bricks if an open ambient is used in the cooling; or (4) the process for upgrading metallurgical-grade silicon for Si PV.

The measurement of carbon in silicon at levels of $5-10 \times 10^{17}/cm^3$ implies a desired detection limit of $5-10 \times 10^{16}/cm^3$. The chosen analytical method can depend upon the form of the silicon to be tested. For example, FTIR can be used for single crystal silicon wafers according to SEMI MF1391-1107 [6]. However, this method has some special challenges for multi-crystalline wafers in that the reference substrate is to be the same thickness as the solar wafer and have a carbon content below 2×10^{15} and the test method is limited to resistivity above 3 ohm-cm for p-type wafers which is higher than Si solar wafers. If the carbon needs to be measured in feedstock bricks, chunks, granules or flakes, FTIR is not appropriate.

SIMS can measure the carbon concentration in all these forms of silicon. SIMS has been used for over 15 years to measure [C] in heavily-doped single crystal electronic-grade silicon, and this method works for all forms of silicon.

Nitrogen

Nitrogen can form nitride-type defects in multi-crystalline PV Si. These defects may affect minority carrier lifetime. But there is another more fundamental reason to measure nitrogen in PV Si. Nitrogen can have an important effect on the Si vacancy and Si interstitial concentrations in silicon [7], and these point defect concentrations affect the diffusion of metal contaminants. The diffusion of metals is important in the segregation of metals in directional solidification furnaces, in gettering schemes (such as phosphorus gettering), and in other processes designed to deactivate metals from being minority carrier lifetime degraders.

The main recognized source of nitrogen in silicon is the nitride release liner in multi-crystalline furnaces, but there may also be nitrogen introduced from the less expensive crucibles used in PV Si. Residual nitrogen may also be present in UMG-Si.

The measurement of nitrogen in silicon at levels of $1 \times 10^{16}/cm^3$ implies a desired detection limit of $1 \times 10^{15}/cm^3$. The chosen analytical method can depend upon the form of the silicon to be tested. For example, FTIR can be used for single crystal silicon wafers but the quantification is complicated by the many forms of nitrogen in silicon [8]. If the nitrogen needs to be measured in feedstock bricks, chunks, granules or flakes, FTIR is not appropriate.

SIMS can measure the nitrogen concentration in all these forms of silicon. SIMS has been used for over 15 years to measure [N] in single crystal electronic-grade silicon as per SEMI MF2139-1103 [9], and this method works for all forms of silicon.

158

SIMS

Secondary Ion Mass Spectrometry (SIMS) [10] analysis provides an accurate quantitative measurement for impurities in PV Si at the sub-6N level (sometimes needing high mass resolution). SIMS can measure all elements of the periodic table, but is especially good for dopants (B, P, As, Sb, Al, In) independent of electrical activity and for atmospherics (H, O, C, N) independent of their chemical state.

During a SIMS analysis, the samples are sputtered by a focused energetic primary ion beam, either oxygen (O_2^+) or cesium (Cs^+). Secondary ions formed during the sputtering process are accelerated away from the sample surface. Secondary ions are energy separated by an electrostatic analyzer and mass separated based on their mass/charge ratio by a magnetic mass analyzer. Oxygen (O_2^+) beam sputtering is used to enhance ion yield of electropositive species (boron and metal elements); cesium (Cs^+) beam sputtering is used to enhance ion yield of electronegative species (P, As, Sb and atmospherics species).

By choosing the proper primary beam and using optimized instrument conditions, SIMS can provide excellent detection limits at sub-ppm to ppt level.

SIMS is both accurate by using traceable reference materials and precise using highly developed protocols and instrumentation.

PV Si can come in many forms, including powders. SIMS can be used to measure wafers, chunks, granules, flakes or even powders if the size of the powder is greater than 300 um.

EXPERIMENTAL DESIGN

In the following case study, SIMS measurements of C, O and N were made on various silicon materials in the process of making Si solar cells. First, two sources of Si feedstock were analyzed: (a) polysilicon from a modified Siemens process, called "7N" Si feedstock here, and (b) upgraded metallurgical grade silicon (UMG-S), called "5N" Si feedstock here. The 7N purity was limited by Cl and H in the bulk. These two feedstocks represent two ways (and purity and cost) for manufacturing the purified Si fed into directional solidification furnaces that melt and regrow the silicon into very large multi-crystalline blocks. After the blocks are grown, they are cut into subsections called bricks from which Si solar wafers are then cut. Lastly, the Si solar wafer is converted into a Si solar cell using various depositions and thermal processing. To save costs sometimes the 7N (higher cost, higher purity) and 5N (lower cost, lower purity) are mixed in some predetermined proportion in the directional solidification furnace.

In this study they were mixed in the following proportion: 7N:5N of 80:20, and 7N:5N of 100:0 (no mixing) in two directional solidification furnaces to make two blocks. For each block 16 bricks were cut out, and solar wafers were sliced from the bricks and polish etched. The slicing was done perpendicular to the solidification direction. Slices from three of the 16 bricks were taken for measurements: corner brick, center brick, and side brick.

SIMS measurements using a Cs^+ primary ion beam were made on the following eight samples:

7N Si feedstock: **7N PV Si**
5N Si feedstock: **5N PV Si**

Wafer from the corner brick of 7N grown block: **7N-Corner**
Wafer from the center brick of 7N grown block: **7N-Center**
Wafer from the edge brick of 7N grown block: **7N-Edge**

Wafer from the corner brick of 7N:5N (80:20) block: **7N:5N-Corner**
Wafer from the center brick of 7N:5N (80:20) block: **7N:5N-Center**
Wafer from the edge brick of 7N:5N (80:20) block: **7N:5N-Edge**

Carbon and oxygen measurements of wafers were done on two separate days months apart and on two separate pieces.

DISCUSSION

The SIMS results are shown in Table I. The two columns for [C] and for [O] represent repeat measurements several months apart.

Table I. SIMS Results on C, O and N Contamination (units of atoms/cc)

	[C]	[C]	[O]	[O]	[N]
7N PV Si	<5E15		<1E16		3.E+14
5N PV Si	3.7E+17		2.0E+18		5.5E+15 N precipitates
Detection Limit	5.E+15		1.E+16		2.E+14
7N-Corner	5.6E+17	5.4E+17	1.1E+17	1.1E+17	1.3E+16
7N-Center	6.3E+17	7.8E+17	7.7E+16	8.2E+16	1.1E+16
7N-Edge	4.8E+17	6.0E+17	8.4E+16	7.1E+16	9.6E+15
Detection Limit	2.E+16	3.E+16	2.E+16	2.E+16	7.E+14
7N:5N Corner	3.9E+17	4.3E+17	1.6E+17	1.5E+17	1.1E+16
7N:5N Center	4.0E+17	5.8E+17	1.6E+17	2.0E+17	1.4E+16
7N:5N Edge	4.4E+17	5.4E+17	1.4E+17	1.2E+17	1.2E+16
Detection Limit	2.E+16	3.E+16	2.E+16	2.E+16	7.E+14

	7N/5N ratioed to 7N				
Comer	0.70	0.80	1.45	1.36	0.85
Center	0.63	0.74	2.08	2.44	1.27
Edge	0.92	0.90	1.67	1.69	1.25

The 7N PV Si (modified Siemens process) has no detectable [C] or [O] below $5 \times 10^{15}/cm^3$ and $1 \times 10^{16}/cm^3$ respectively. This is as expected for a Siemens-based polysilicon process. The 7N PV Si does have detectable [N] which is about an order of magnitude higher than found in electronic grade Si substrates [11].

The 5N PV Si which is an upgraded metallurgical grade silicon (UMG-Si) does have much higher levels of C, O and N, and there are nitrogen precipitates, presumably silicon nitride. We have tested other UMG-Si which was cooled in non-silicon nitride forms and not detected nitrogen above $1 \times 10^{15}/cm^3$, so this N may come from a furnace liner. The carbon levels in the 5N PV Si are below the SEMI PV Si wafer specification for mc-Si wafers, but the oxygen level is not.

The directional solidification of 7N into blocks for Si wafers adds C, O and N far above the original 7N feedstock, but the C and O are still less than the SEMI PV Si solar wafer specification. The 7N-Edge brick has slightly less C, O and N than the 7N-Center or 7N-Corner bricks. (NOTE: the 7N-Edge brick was cropped from the original 7N block.) There is no SEMI specification for nitrogen in PV Si wafers, but the $\sim 1 \times 10^{16}/cm^3$ [N] level is high and presumably comes from the silicon nitride liner in the directional solidification furnace, but this level of nitride is about the same for all three bricks which are located in difference places of the block. This suggests that from the standpoint of C, O and N, the higher purity C, O and N in 7N PV Si is overkill, i.e. the subsequent processing adds much higher levels of C, O and N.

What happens then if the 7N and 5N are mixed 80/20? Compared to the pure 7N case, the carbon is reduced but the oxygen and nitrogen are increased. Note that we do not have a 5N control. The block grown from 7N feedstock and the block grown from 7N/5N feedstock were grown in different directional solidification furnaces. The unexpected apparent C reduction from mixing pure 7N with dirty 5N feedstock may be due to differences in the C contamination from the different furnaces.

CONCLUSIONS

The conclusion of this study twofold: (a) the primary opportunity to reduce C, O and N contamination in mc-Si solar cells is at the directional solidification process, and (b) the costly specification of highly pure Si feedstock is unnecessary *from a C, O and N perspective* if a directional solidification process is used.

This study does not draw any conclusions on other elements, such as dopants and metals.

REFERENCES

1. M. Ghosh, A. Muller, D. Sontag. H. Neuhaus, K. Bothe, and J. Schmidt, 21st PV SEC, Dresden, p. 1326 (2006).
2. Specification for Silicon Wafers for Use as Photovoltaic Solar Cells, SEMI International Standard M6-0707, (published by SEMI, San Jose, CA, www.semi.org)
3. Test Method for Interstitial Oxygen Content of Silicon by Infrared Absorption with Short Baseline, SEMI International Standard MF1188-1105, ibid.
4. Test Method for Measuring Oxygen Concentration in Heavily Doped Silicon Substrates by Secondary Ion Mass Spectrometry, SEMI International Standard MF1366-1107, ibid.

5. S. Riepe, I. E. Reis, and W. Koch, "Solar Silicon Material Research Network SOLARFOCUS (Solarsilizium Forschungscluster)," 23rd PV SEC, (2008)
6. Test Method for Subsitutional Atomic Carbon Content of Silicon by Infrared Absorption, SEMI International Standard MF1391-1107, ibid.
7. Milind S. Kulkarni, J. Crystal Growth **310**, 324 (2008).
8. US Patent 6803576 - Analytical method to measure nitrogen concentration in single crystal silicon, October 12, 2004, (and references therein).
9. Test Method for Measuring Nitrogen Concentration in Silicon Substrates by Secondary Ion Mass Spectrometry, SEMI International Standard MF2139-1103, ibid.
10. *Secondary Ion Mass Spectrometry, A Practical Handbook for Depth Profiling and Bulk Impurity Analysis*, edited by Robert G. Wilson, Fred A. Stevie, and Charles W. Magee, published by John Wiley & Sons, New York (1989).
11. L. Wang, Evans Analytical Group (private communication)

New Technologies

Mater. Res. Soc. Symp. Proc. Vol. 1123 © 2009 Materials Research Society   1123-P04-08

A Novel Solution Processable Electron Acceptor, $C_{60}(CN)_2$, for Bulk Heterojunction Photovoltaic Applications

Vaishali R. Koppolu[1], Mool C. Gupta[1], Will Bagienski[1], Yang Shen[1], Chunying Shu[2], Harry W. Gibson[2], Harry C. Dorn[2]

[1] Charles L. Brown Department of Electrical and Computer Engineering, 351 McCormick Road, University of Virginia, Charlottesville, VA 22904, U.S.A.

[2] Department of Chemistry, 107 Davidson Hall, Virginia Polytechnic Institute and State University, Blacksburg, VA 24061, U.S.A.

ABSTRACT

Photovoltaic devices based on soluble conjugated polymers have gained great interest in recent years because of the potential low cost of production and ease of fabrication. PCBM ([6,6]-phenyl-C_{61}-butyric acid methyl ester), a fullerene derivative, has been extensively investigated as a solution processable electron acceptor for bulk-heterojunction (BHJ) photovoltaic devices blended with conjugate polymers like P3HT [poly(3-hexylthiophene)]. Here, we investigated a novel solution processable organic semiconductor, $C_{60}(CN)_2$, as an electron acceptor for bulk heterojunction photovoltaic applications. Optical and electrical properties of $C_{60}(CN)_2$ are studied and compared with PCBM. Blend devices with P3HT and $C_{60}(CN)_2$ have been fabricated and compared with P3HT-PCBM devices. The effect of thermal annealing on the device performance is evaluated. Open circuit voltage, short circuit current, fill factor and total efficiency data are compared with PCBM based devices.

INTRODUCTION

Bulk heterojunction (BHJ) solar cells fabricated by blending conjugated polymers with fullerenes have resulted in great improvements in the energy conversion efficiencies in recent years. The most commonly and extensively used materials for the fabrication of BJHs are P3HT [poly(3-hexylthiophene)] and PCBM ([6,6]-phenyl-C_{61}-butyric acid methyl ester). Various studies have shown that the device performance is strongly related to nanomorphology of blends, processing conditions, solvents used and on thermal treatment [1,2]. With optimized device structure and fabrication conditions, P3HT:PCBM-based BHJ solar cells can generate efficiencies up to ~5% [1,2]. At present substituted poly(p-phenylene vinylene)s (PPVs) and polythiophenes (PThs) are typically used as donors and fullerene derivatives primarily are used as acceptors for these photovoltaic devices [3]. For further improvement of organic photovoltaic power conversion efficiencies, new donor-acceptor materials need to be examined. Here we investigated a solution processable organic semiconductor, $C_{60}(CN)_2$, as a novel electron acceptor for BHJ applications. Cyclic voltammetry has shown that $C_{60}(CN)_2$ is a stronger electron acceptor than the parent C_{60} [4].

Although this molecule was first synthesized by Keshavaraz-K et al. in 1995 [5], its electrical properties and photovoltaic applications were never investigated. In this study we investigated $C_{60}(CN)_2$ as a solution processable electron acceptor for the first time for BJH photovoltaic applications. Solar cells based on a blend of P3HT as donor and $C_{60}(CN)_2$ as

---

[1] Corresponding author E-mail: mgupta@virginia.edu

acceptor were fabricated and their performance is compared with P3HT:PCBM devices. All the devices were fabricated in air atmosphere. Thermal treatment of the blend devices has shown significant improvement in the device performance. Molecular structures of the active materials: P3HT, PCBM and $C_{60}(CN)_2$ are shown in Figure 1. Device structure and an energy-level diagram showing the HOMO and LUMO energies of each of the component materials are shown in Figure 2.

P3HT   PCBM $C_{60}(CN)_2$

Figure 1. Molecular structures of active materials: P3HT, PCBM and $C_{60}(CN)_2$

(a)

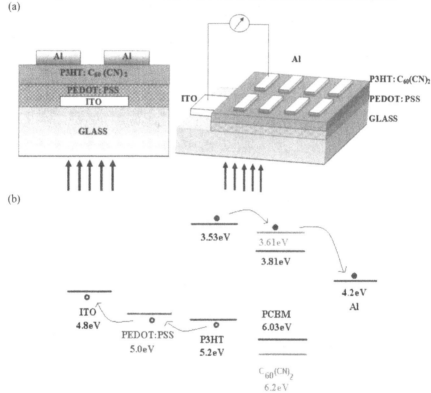

(b)

Figure 2. (a) Device structure- cross-sectional view and top view and (b) Energy level diagram showing HOMO and LUMO levels of each of the component materials

166

EXPERIMENTAL

Materials

$C_{60}(CN)_2$ was synthesized and purified according to literature procedures [5,6]. ITO glass substrates (R_s = 8-12 Ω per square) were purchased from Delta Technologies Ltd. PEDOT: PSS (Clevios P VP AI 4083) was purchased from H. C. Starck and regioregular P3HT (M_n ~64,000) was purchased from Sigma-Aldrich. PCBM was purchased from Nano-C.

Photovoltaic Device Fabrication

The photovoltaic devices (Figure 1) were fabricated using a blend of P3HT:PCBM and P3HT:$C_{60}(CN)_2$ in the ratio of 45:55 wt. % in chlorobenzene that was stirred overnight in air. Indium-tin-oxide (ITO) coated glass was cleaned with deionized water, acetone and reagent alcohol sequentially and dried with N_2. The ITO surface was then modified by spin coating a thin layer (~40 nm) of PEDOT:PSS at 4000 rpm for 60 seconds. This layer was soft baked on a hot plate at 110 °C for 15 mins in air. The active layer (~130-160 nm) was then spin coated at 500 rpm for 30 seconds. The devices were then dried in air for 20 mins in covered glass petri dishes to allow the wet films to solidify [7]. The devices were then soft baked in an oven at 60 °C for 30 mins. An 80 nm Al film was evaporated in high vacuum (10^{-6} torr) using a shadow mask. The active device area was 0.12 – 0.14 cm^2. All the fabrication procedures were carried out in air.

Characterization

Absorption spectra for the pristine films and blended films coated on quartz were obtained using Perkin Elmer EZ201Spectrophotometer. The current–voltage characteristics were measured using a Keithley 2611 source meter, custom tungsten lamp with custom filter built by PV Measurements, Inc. and a BK Precision power supply 1690 EXD DC. For white light efficiency measurements at 100 mW/cm^2, an Oriel solar light simulator with AM 1.5 filter was used. For quantum efficiency measurements, different wavelengths of light were selected with a Jobin Yvon 270M monochromator. A Newport model 69901 Xe lamp with 69911 power supply was used as the light source. The light intensity was measured using a Newport 842-PE meter with 918D-UV-OD3 detector.

RESULTS AND DISCUSSION

Figure 3 shows the UV- Vis absorption spectra measured for pure P3HT, PCBM and $C_{60}(CN)_2$ films coated on quartz and for P3HT:PCBM and P3HT:$C_{60}(CN)_2$ composite films. The solid lines indicate the spectra for untreated films and the dotted lines indicate the spectra after the films were annealed at 130 °C for 2 mins in air. From the spectra it can be seen that thermal annealing of the films has no effect on the absorption for pure materials. The spectra for pure PCBM and $C_{60}(CN)_2$ also indicate that for the same concentration, $C_{60}(CN)_2$ has a lower absorption in the UV range and higher absorption in the visible region compared to PCBM. Comparing the spectra of P3HT:PCBM and P3HT:$C_{60}(CN)_2$ as shown in Figure 3 (b), the absorption peaks appear at same locations (520, 565 and 605 nm) for the annealed films. For untreated films, the spectrum in the visible region for the P3HT:$C_{60}(CN)_2$ system exhibits a blue shift with lower absorption as compared to P3HT:PCBM system. Since the absorption spectrum

167

of the blends for the unannealed and annealed films is comparable, the device performance was evaluated by fabricating photovoltaic cells utilizing P3HT:PCBM and P3HT:$C_{60}(CN)_2$ blends as the active layers. Post-thermal annealing was carried out on a hot plate at 130 °C for 2 mins in air.

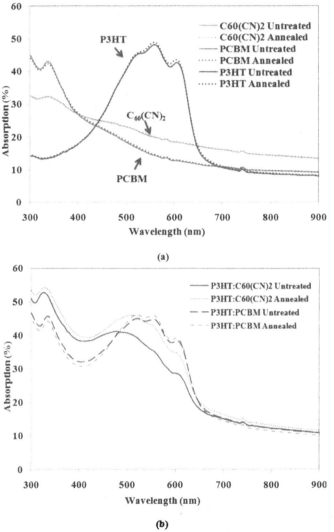

(a)

(b)

Figure 3. UV- Vis absorption spectra for (a) pure P3HT, PCBM and $C_{60}(CN)_2$ (b) P3HT:PCBM and P3HT:$C_{60}(CN)_2$ composite films

The current density-voltage (J-V) characteristics of the cells in the dark and under simulated AM 1.5 solar illumination are shown in Figure 4, and the results are summarized in Table 1. The maximum power conversion efficiency of the thermally annealed P3HT:$C_{60}(CN)_2$ device was 0.6%, while for the thermally annealed P3HT:PCBM device it was 2.7%. Post-thermally treated P3HT:$C_{60}(CN)_2$ blend devices yielded much higher performance as indicated in Table 1.

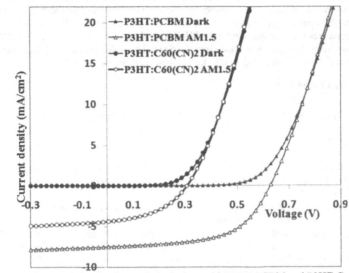

Figure 4. J- V characteristics for photovoltaic devices with P3HT:PCBM and P3HT:$C_{60}(CN)_2$ systems

Table 1. Device performance summary for P3HT:PCBM and P3HT:$C_{60}(CN)_2$ blends

Material System	Voc (V)	J_{sc} (mA/cm^2)	Fill Factor (%)	η_e (%)
P3HT:$C_{60}(CN)_2$ No Annealing	0.17	0.097	24.05	4 x10^{-3}
P3HT:PCBM Annealed-130 °C/2 min	0.63	7.47	57.12	2.7
P3HT:$C_{60}(CN)_2$ Annealed – 130 °C/2 min	0.31	4.39	43.96	0.6

CONCLUSIONS

A novel solution processable electron acceptor, $C_{60}(CN)_2$, was explored for BHJ photovoltaic applications by blending it with P3HT. A comparison of absorption spectra of pure materials and blends was carried out and the effect of thermal annealing on the absorption and hence the device performance was evaluated. The factors limiting the efficiency of P3HT:C_{60}(CN)$_2$ devices are lower short circuit current, fill factor, and open circuit voltage as compared to P3HT:PCBM.

All the experiments were carried out in air atmosphere which explains lower efficiency than obtained in a controlled environment. Further improvements in device performance can be achieved by control of morphology, proper blending ratio with P3HT and thermal annealing conditions. The open circuit voltage, $V_{oc,}$ was about half that of P3HT:PCBM material system which necessitates accurate measurement of HOMO and LUMO levels for C_{60}(CN)$_2$ materials using ultraviolet photoelectron spectroscopy (UPS) method. Also the purity of the material can be further improved which would lead to further enhancement in efficiency.

ACKNOWLEDGMENTS

We thank National Science Foundation for financial support.

REFERENCES

1. A. Pivrikas, P. Stadler, H. Neugebauer, N. S. Sariciftci, "Substituting the post production treatment for bulk-heterojunction solar cells using chemical additives", *Organic Electronics* **9**, 775-782 (2008)
2. W.-Y. Wong, X.-Z. Wang, Z. He, A. B. Djurisic, C.-T. Yip, K.-Y. Cheung, H. Wang, C. A. K. Mak and W. K. Chan, *Nature Materials*, **6** 521-527 (2007)
3. S. Lu, Solar cells based on organic materials may provide low-cost power, *SPIE* (2007)
4. Yoshida Yukihiro, Otsuka Akihiro, Saito Gunzi, *Mol. Cryst. Liq. Cryst.*, **376**, 189-196 (2002)
5. M. Keshavaraz-K, B. Knight, G. Srdanov and F. Wudl, *J. Am. Chem. Soc.*, **117**, 11371-11372 (1995)
6. Y. Yoshida, A. Otsuka, O. O. Drazdova, K. Yakushi and G. Saito, *J. Mater. Chem.*, **13**, 252-257 (2003)
7. G. Li, V. Shrotriya, J. Huang, Y. Yao, T. Moriarty, K. Emery and Y. Yang, *Nature Materials*, **4** 864-868 (2005)

Mater. Res. Soc. Symp. Proc. Vol. 1123 © 2009 Materials Research Society   1123-P05-08

Thickness Dependent Effects of Thermal Annealing and Solvent Vapor Treatment of Poly (3-hexylthiophene) and Fullerene Bulk Heterojunction Photovoltaics

Zhouying Zhao[1], Lynn Rice[1], Harry Efstathiadis[1] and Pradeep Haldar[1*]
[1]College of Nanoscale Science and Engineering, University at Albany, State University of New York, 255 Fuller Road, Albany, NY 12203, U.S.A.

ABSTRACT

We have utilized room-temperature solvent vapor treatment followed by thermal annealing to process bulk heterojunction (BHJ) photovoltaic devices based on blends of poly (3-hexylthiophene) (P3HT) and phenyl-C61-butyric acid methyl ester (PCBM) of varied active layer thickness. The morphological and photovoltaic performance characteristics of the cells subject to these treatments were found to be dependent on active layer thickness. The devices were characterized using atomic force microscopy (AFM) and opto-electrical and external quantum efficiency measurements in order to analyze the mechanism underlying the observed trend. Performance indicators including fill factor, short-circuit current and power conversion efficiency were correlated to the ordering of device active layers and morphology. The maximum power conversion efficiency achieved was 4.1 %.

INTRODUCTION

From the earliest work [1] onward, and particularly with the advent of bulk heterojunction (BHJ) as opposed to bilayer processing [2], the P3HT/PCBM system has emerged as a model system for organic photovoltaic devices because of the system's balanced charge transfer and high absorption in the visible range. In comparison to silicon-based photovoltaics, organic PV devices are easier to process and potentially more economical [3]. To date, record power conversion efficiency of over 6 percent has been achieved by optimizing processing in the P3HT/PCBM system [4].

The effect of active layer annealing in BHJ photovoltaic devices based on blends of these materials is of special interest because the ultimate cell performance and efficiency bear a strong dependence on the annealing method and conditions.

Optimizations that take annealing into consideration have been performed a number of times on the P3HT/PCBM heterojunction system, resulting in state-of-the-art power conversion efficiencies. [5-8]. Features seen upon thermal annealing of these BHJ devices include increased photon absorption, reduced series resistance of the bulk blend film, and better contacts with the cathode and anode [5].

The addition of PCBM to P3HT disrupts P3HT ordering and annealing facilitates phase separation and restores the polymer to its pristine state of ordered structure [9, 10]. Annealing increases the P3HT molecule chains' π-π overlap, resulting in a red shift of the absorption maxima [11]. Higher crystallinity of P3HT and more fluidic mobility of fullerenes in the BHJ mesophase resulting in the fullerenes' uniform dispersion within the layer dramatically effects PV performance [12]. With ordering and formation of PCBM aggregates comes marked differences in charge carrier mobility and solar energy absorption which improves the short circuit current (I_{sc}), fill factor (FF) and open circuit voltage (V_{oc}) in annealed devices [3, 10, 13]. Explicitly, a modification in hopping mechanisms seen in annealed BHJ systems results in lower hopping activation energies because of improved charge carrier mobility in the active layer [14].

With respect to processing, Y. Kim et al noted that the solvent plays a role in modifying film morphology, finding that active layers cast from chlorobenzene demonstrated more ordered mesophase morphology and ultimately performed more efficiently than those cast from dichlorobenzene [15].

Their work concluded that the observed preferential solvent effects were rooted in the different evaporation speed of the active layer during processing. Their conclusion was corroborated by experiments in which the only variable was evaporation speed [16]. In fact, it was also observed that, for certain evaporation speeds, no thermal anneal was necessary to restore P3HT crystallinity.

Photocurrent and mobility measurements performed on slowly and quickly evaporated layers indicate that hole mobility is reduced and hole transport is more dispersive in quickly evaporated layers (though there is a higher electron mobility), while, in slowly evaporated layers, hole transport is more non-dispersive and enhanced [17].

The ability to create nanocrystalline P3HT/PCBM domains absent heat treatment followed the advent of the recent solvent vapor treatment, in which the BHJ cells are subjected to solvent vapors at room temperature. Solvent vapor treatment has been observed to have a similar effect as thermal annealing, with solvent vapor-treated samples ultimately achieving higher power conversion efficiencies than their thermally-annealed counterparts, in part due to the fact that they are not subject to thermal stress [19, 20].

Even more recently, researchers found that devices subjected to a combination of solvent vapor treatment and thermal annealing perform better than otherwise identically processed devices subjected to either process alone. They propose that solvent vapor treatment activates self-organization of the P3HT molecule chains, while thermal annealing activates diffusion and aggregation of PCBM [21]. Still, an understanding of the precise mechanism behind annealing and corresponding interpenetrating of P3HT/PCBM BHJ, and recorded relationships governing annealing process conditions and their direct effect on cell performance, remains to be developed.

We have investigated the effect of solvent treatment and thermal annealing processes and found these effects to be specific to active layer thickness. In this work, we performed solvent and thermal annealing on samples of various thicknesses and found that, not only is there is a systematic change in performance upon device annealing that is contingent on active layer thickness, but that, at some critical thickness, an inflection point is observed where the effect of thermal annealing switches from improving device performance to degrading device performance.

EXPERIMENT

Our photovoltaic devices were fabricated on commercially pre-patterned indium tin oxide coated glass slides. The substrates were pre-cleaned in an ultrasonic bath of deionized water and then treated in UV-Ozone using Jelight UVO cleaner. Polyethylenedioxythiophene- poly styrene sulfonate (PEDOT-PSS) (Baytron P) was spin-coated onto the substrates and annealed at 170° C for 4 min to form a 60 nm -thick film.

Prior to cell fabrication, a solution consisting of 17 mg/mL regioregular P3HT (from Rieke Metals, Inc.) in 1,2-dichlorobenzene was added to PCBM (from American Dye Source, Inc) in a 1:0.6 weight ratio and then stirred at 45 °C for 36 h in a light-protected environment in nitrogen ambient.

The substrates were transferred inside a nitrogen-filled glove box where an active layer was spin coated from the P3HT/PCBM solution at speeds of 500, 1500 and 2000 rpm for 60 s. Each of the substrates was transferred to a plastic container and exposed to vapor of the residual solvent. A ~1 nm thick layer of cesium fluoride followed by a 40 nm thick top contact of aluminum were thermally evaporated onto the active layer in a vacuum chamber located inside the glove box.

The as-prepared cell performance was measured under AM 1.5 illumination using a halogen lamp solar simulator in a nitrogen environment at room temperature. The lamp had been calibrated to 100 mW/ cm^2 with a Si photodiode (Edmund optics) coupled to an integrating sphere. Current-voltage measurements were collected with a Keithley 237 source measurements unit. The devices were then annealed at 160 ° C for 8 min and their photovoltaic performance was again tested under the same conditions. External quantum efficiency (EQE) measurements of the annealed cells were performed using an Oriel Instruments 1000 W Xenon lamp with a monochromator and a Newport power meter outside the glove box after the device area was encapsulated with an epoxy-like layer. Film surface morphology was measured by tapping mode AFM using a Digital Instruments Nanoscope. The thickness measurements were conducted on an Alpha step profilometer and confirmed by scanning electron microscopy. The quantum efficiency was obtained from short circuit current and light intensity of monochromic wavelength illumination. Four to six devices were measured for each type or set of samples that have an active area per device of 3 mm^2.

RESULTS AND DISCUSSION

We have fabricated three sets of photovoltaic devices with active layer thicknesses of 100 nm, 160 nm and 200 nm (with an accuracy of +/- 5 nm), corresponding to spin casting speeds of 1500 rpm, 1000 rpm and 500 rpm, respectively. Each set of devices was then subjected to a degree of specific solvent anneal (solvent vapor treatment) depending on the spin speed used and residual solvent left, followed by an identical thermal anneal. Typical diode characteristic curves were obtained before and after thermal annealing for all devices fabricated. Figure 1 compares

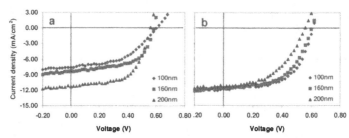

Figure 1. Light I-V curves for P3HT/PCBM heterojunction photovotaic cells with different active layer thickness as marked in the figure (a) solvent annealing only and (b) solvent/thermal annealing.

the current versus voltage (I-V) characteristics under AM1.5 illumination from the devices receiving solvent vapor treatment only and solvent/thermal annealing, respectively. As seen from the figure, the best photovoltaic performance was initially obtained from the 200 nm device with solvent vapor treatment (Figure 1 (a)) but subsequently obtained from the 100 nm device with followed thermal annealing (Figure 1 (b)). This trend is repeated for the other devices of the same active layer thickness as seen from the power conversion efficiency variations with the annealing treatments depicted in Figure 2.

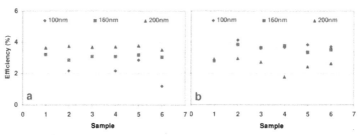

Figure 2. Effect of annealing treatment on power conversion efficiency of P3HT/PCBM heterojunction photovoltaic cells of different active layer thickness: (a) solvent annealing only and (b) solvent/thermal annealing. With solvent annealing, the 200 nm-thick cell exhibits the highest performance, while after solvent/thermal annealing, the 100 nm-thick cell exhibits the best performance.

The statistics of the overall performances were obtained for the devices described above. Before thermal annealing or after solvent vapor treatment, the 100 nm devices had an average power conversion efficiency of 2.1 %, average FF of 0.46 and average V_{oc} of 0.58 V. After thermal annealing, the cell performance parameters improved to average efficiency of 3.7 %, FF of 0.55 and V_{oc} of 0.6 V, while the best performing device exhibited an efficiency of 4.1%.

The solvent vapor treated 160 nm devices showed 3.1 % efficiency and had FF of 0.62 and V_{oc} of 0.57 V. After thermal annealing, the average FF of these devices dropped to 0.52, while the V_{oc} increased to 0.59 and efficiency increased to 3.5 %.

174

The 200 nm thick devices had 3.5% efficient after solvent annealing, with average FF of 0.58 and V_{oc} of 0.56. Upon thermal annealing, all of these parameters decreased, resulting in 2.5% average efficiency, FF of 0.43 and V_{oc} of 0.55.

Generally, the purpose of annealing PCBM/P3HT BHJ devices is to restore or allow the crystallization of P3HT that is disrupted in the presence of PCBM during device processing [9-14]. While both thermal annealing and solvent vapor treatment techniques improve P3HT ordering, the mechanism by which each technique does so is different, and hence device conditions before thermal annealing have a significant impact on the ultimate effect of the annealing on developed cell performance.

Within the range of active layer thickness investigated, our results indicate that the annealing effect on power conversion efficiency is dependent on film thickness. While thermal annealing was shown to generally improve power conversion efficiency in optimized devices [4], we present a set of devices with relatively thick active layers that don't conform to this broad trend; their performance is systematically lowered upon thermal annealing after solvent vapor treatment.

Here, we make the distinction between "solvent vapor treatment" in which the active layer is not completely dried during the spin coating process and the remainder of the solvent is let to evaporate slowly (what otherwise might be termed "solvent evaporation") and the aforementioned solvent-vapor annealing, in which spin cast, incompletely dried active layers are subject to vapors of solvents such as dichlorobenzene or chlorobenzene to complete phase separation and polymer ordering.

For the thinnest sample (100 nm), the average power conversion efficiency of devices increased from 2.1% to 3.7% in going from the initial solvent treatment step through thermal annealing. For the intermediate thickness sample (160 nm), the average power conversion efficiency of devices did modestly improve from 3.1% up to 3.5% upon thermal annealing. In contrast, for the thickest sample of 200 nm, the average power conversion efficiency after thermal annealing decreased from 3.5% to 2.5%. It is notable that devices identical in every processing aspect except thickness of the active layer had opposite responses to thermal annealing. In thin films less than 160 nm, annealing improved active layer crystallization and networking, while in thick films up to 200 nm, networking was degraded in this thermal anneal step.

The reason for this observed difference in annealing effect stems from the fact that thermal and solvent annealing is mechanistically different. The dominant mechanism of solvent annealing is self assembly of P3HT molecular chains. The thickest films are relatively rich in solvent (and are visibly wet) when undergoing slow spin casting, and hence the P3HT molecules are free to align themselves in this liquid layer and respond dramatically to solvent annealing. The device microstructure domain thereby undergoes significant changes during the solvent annealing of thicker, wetter films [15].

Thinner films have had most of the solvent evaporated in the spin coating step, forcing the P3HT chains into a more rigid phase and little solvent is available to produce an annealing effect on the performance of these devices because most of the solvent had already been evaporated prior to the treatment.

The fragile self organization that is initiated in the solvent annealing step in the thicker films is disrupted by thermal annealing, and the microstructure formed is not maintained. Parts of the networking developed in the solvent evaporation are destroyed by thermal stress [19, 20]. In fact, for thick film devices, this effect of disrupted networking has a stronger influence on device

175

performance than the positive effect developed in thermal annealing of improving charge transport in the active layer, accounting for a net performance degradation observed upon thermal annealing.

Thinner films, on the other hand, do not receive as much self-organization in the absence of residual solvent and vapor, and hence phase separation and polymer ordering hindered in spin process is not restored effectively. The primary effect observed during thermal annealing is the improved diffusion and aggregation of PCBM clusters and polymer ordering.

The EQE measurements for the thermal annealed cells are plotted in Figure 3. The device with a 100 nm-thick active layer shows a maximum EQE of 67% at 500nm wavelength, a ~18% improvement over the device with a 160 nm-thick active layer. While thicker films still absorb the most light after thermal annealing, as a result of well ordered P3HT [11], the absorbance of light does not correspond to improved efficiency. In fact, the opposite trend is observed: the best performance is achieved in the 100 nm active layer devices. This trend suggests that both microstructure and polymer ordering in the active layer are critical to device performance and that favorable morphology is more efficiently achieved in the thin film realm after thermal annealing.

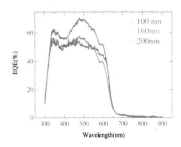

Figure 3. External quantum efficiency measurements for P3HT/PCBM heterojunction photovoltaic cells with different active layer thickness after solvent/thermal annealing. The device with a 100 nm-thick active layer shows maximum EQE of 67% at wavelength of ~500nm.

In these experiments, surface roughness was not found to be a critical factor in device performance. Atomic Force Microscopy yields surface roughnesses of 1.3 nm, 1.2 nm and 4.3 nm RMS for the 100, 160 and 200 nm active layers, respectively (Figure 4). The 100 and 160 nm films were comparably rough, while the 200 nm film was significantly rougher. Several groups in the past have noted enhanced device efficiency with greater surface roughness [5], but no such effect was observed in this work. If the effect is present, it is overshadowed by the efficiency drop that resulted from the altered shift in microstructure domains that resulted from the thermal annealing step.

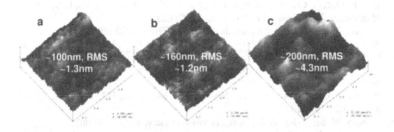

Figure 4. AFM surface morphology of P3HT/PCBM heterojunction photovoltaic cells after solvent/thermal annealing. The film thickness and surface roughness are marked on image (a), (b) and (c).

Understanding the mechanisms by which thermal and solvent annealing take place is paramount to continuing to improve device efficiencies. Ultimately, if thermal and solvent annealing techniques were optimized for thick-layer devices, the same polymer ordering (and corresponding charge carrier mobility and light absorption) and networking that exists in the thin-film realm could be achieved. Thus, current thickness limitations on performance in organic photovoltaic devices could be overcome and optimal thicker devices could be viably processed. Such a breakthrough would enable devices to absorb more light and generate more power, and represent a major step towards the development of photovoltaic devices.

CONCLUSIONS

We have shown that solvent vapor treatment improves the performance of 200 nm-thick layer devices more noticeably than thin-layer devices. Followed thermal annealing was found to be most beneficial in improving power efficiency of devices with active layer thicknesses in the range of 100 to 160 nm. We believe that this is due to a combined effect of polymer ordering and optimization of morphology/percolation in the blend film, which varies with film thickness and is sensitive to annealing treatments.

ACKNOWLEDGMENTS

The authors would like to thank Dr. George G. Malliaras and his research group at Cornell University for support of partial experiments. Funding for this project was provided by a grant from the National Aeronautics and Space Administration.

REFERENCES

1. Saricifti, N.S., Smilowitz, L. Heeger, A.J., Wudl, F. Science 258, 5987, 1474 (1992).

2. C. J. Brabec, A. Cravino, D. Meissner, N.S. Sariciftci, M.T. Rispens, L. Sanchez, J.C. Hummelen. Thin Solid Films 403-403, 368 (2002).
3. Brabec, C.J. Solar Energy Materials and Solar Cells 83, 273 (2004).
4. K. Kim, J. Liu, M. Namboothiry, D. Carroll. Applied Phys. Letters 90, 163511 (2007).
5. G. Li, Y. Yang J. Appl. Phys. 98, 043704 (2005).
6. T. Erb, U. Zhokhavets, H. Hope, G. Gobsch, M. Al-Ibrahim, O. Ambacher. Ab Thin Solid Films 511-512, 483 (2006).
7. Y. Kim, S. Cook, S.A. Choulis, J. Nelson, J.R. Durrant and D.D.C. Bradley. Synthetic Metals 152, 105 (2005).
8. H. Kim, W. So, S. Moon. J. Korean Physical Society, 48, 441 (2006).
9. T. Aernouts, P. Vanlacke, I. Haeldermans, J. D. Haen, P. Haremans, J. Poortmans, J. Mance. Mater. Res. Soc. Symp. Proc. Vol. 1013 (2007).
10. M. Reyes-Reyes, K. Kim, D. Carroll. Appl. Phys. Lett. 87, 083506 (2005).
11. K. Yazawa, Y. Inoue, T. Yamamoto, N. Askawa, Phys. Rev. B 74, 094204 (2006).
12. M.R. Reyes, K. Kim, J. Dewald, R.L. Sandoval, A. Avadhanula, S. Curran, and D. L. Carroll. Org. Lett. 7, 5749 (2005).
13. P. Vanlaeke, A. Swinnen, I. Haeldermans, G. Vanhoyland, T. Aernouts, D. Cheyns, C. Deibel, J. D¡¦Haen, P. Heremans, J. Poortmans and J.V. Manca Solar Energy Materials & Solar Cells 90, 2150 (2006).
14. K. Kim, J. Liu, M. Namboothiry, D. Carroll. Appl. Phys. Lett. 90, 163511 (2007).
15. Y. Kim, S. Cook, S.A. Choulis, J. Nelson, J.R. Durrant, D.D.C. Bradley. Appl. Phys. Lett. 86, 063502 (2005).
16. P. Vanlaeke, G. Vanhoyland, T. Aernouts, D. Cheyns, C. Deibel, J. Manca P. Heremans, J. Poortmans. Thin Solid Films 511-512, 358 (2006).
17. J. Huang, G. Li, Y. Yang. Appl. Phys. Lett. 87, 1121 05 (2005).
18. W. Ma, C. Yang, X. Gong, K. Lee, A. J. Heeger. Adv. Func Mater, 13, 85 (2003).
19. S. Miller*, G. Fanchini, Y. Lin, C. Li, C. Chen, W. Su and M. Chhowalla. J Mater Chem, 18, 306 (2008).
20. G. Li, Y. Yao, H. Yang, V. Shrotriya, G. Yang, Y. Yang Adv. Funct. Mater., 17, 1636 (2007).
21. Yun Zhao, Zhiyuan Xie,a_ Yao Qu, Yanhou Geng, and Lixiang Wang. Appl. Phys. Lett. 90, 043504 (2007).

179